Statistical Methods for
Fuzzy Data

Statistical Methods for Fuzzy Data

Reinhard Viertl

Vienna University of Technology, Austria

A John Wiley and Sons, Ltd., Publication

This edition first published 2011
© 2011 John Wiley & Sons, Ltd

Registered office
John Wiley & Sons Ltd, The Atrium, Southern Gate, Chichester, West Sussex, PO19 8SQ, United Kingdom

For details of our global editorial offices, for customer services and for information about how to apply for
permission to reuse the copyright material in this book please see our website at www.wiley.com.

Library of Congress Cataloging-in-Publication Data

Viertl, R. (Reinhard)
    Statistical methods for fuzzy data / Reinhard Viertl.
        p. cm.
    Includes bibliographical references and index.
    ISBN 978-0-470-69945-4 (cloth)
    1. Fuzzy measure theory. 2. Fuzzy sets. 3. Mathematical statistics. I. Title.
QA312.5.V54 2010
515'.42–dc22

2010031105

A catalogue record for this book is available from the British Library.

Print ISBN: 978-0-470-69945-4
ePDF ISBN: 978-0-470-97442-1
oBook ISBN: 978-0-470-97441-4
ePub ISBN: 978-0-470-97456-8

Typeset in 10/12pt Times by Aptara Inc., New Delhi, India
Printed and bound in Singapore by Markono Print Media Pte Ltd

Contents

**Part V   BAYESIAN INFERENCE AND FUZZY
          INFORMATION   87**

Preface

Statistics is concerned with the analysis of data and estimation of probability distribution and stochastic models. Therefore the quantitative description of data is essential for statistics.

In standard statistics data are assumed to be numbers, vectors or classical functions. But in applications real data are frequently not precise numbers or vectors, but often more or less *imprecise*, also called *fuzzy*. It is important to note that this kind of uncertainty is different from errors; it is the imprecision of individual observations or measurements.

Whereas counting data can be precise, possibly biased by errors, measurement data of continuous quantities like length, time, volume, concentrations of poisons, amounts of chemicals released to the environment and others, are always not precise real numbers but connected with imprecision.

In measurement analysis usually statistical models are used to describe data uncertainty. But statistical models are describing variability and not the imprecision of individual measurement results. Therefore other models are necessary to quantify the imprecision of measurement results.

For a special kind of data, e.g. data from digital instruments, interval arithmetic can be used to describe the propagation of data imprecision in statistical inference. But there are data of a more general form than intervals, e.g. data obtained from analog instruments or data from oscillographs, or graphical data like color intensity pictures. Therefore it is necessary to have a more general numerical model to describe measurement data.

The most up-to-date concept for this is special fuzzy subsets of the set $\mathbb{R}$ of real numbers, or special fuzzy subsets of the k-dimensional Euclidean space $\mathbb{R}^k$ in the case of vector quantities. These special fuzzy subsets of $\mathbb{R}$ are called nonprecise numbers in the one-dimensional case and nonprecise vectors in the k-dimensional case for $k > 1$. Nonprecise numbers are defined by so-called *characterizing functions* and nonprecise vectors by so-called *vector-characterizing functions*. These are generalizations of *indicator functions* of classical sets in standard set theory. The concept of fuzzy numbers from fuzzy set theory is too restrictive to describe real data. Therefore *nonprecise numbers* are introduced.

By the necessity of the quantitative description of fuzzy data it is necessary to adapt statistical methods to the situation of fuzzy data. This is possible and generalized statistical procedures for fuzzy data are described in this book.

There are also other approaches for the analysis of fuzzy data. Here an approach from the viewpoint of applications is used. Other approaches are mentioned in Appendix A4.

Besides fuzziness of data there is also fuzziness of a priori distributions in Bayesian statistics. So called *fuzzy probability distributions* can be used to model nonprecise a priori knowledge concerning parameters in statistical models.

In the text the necessary foundations of fuzzy models are explained and basic statistical analysis methods for fuzzy samples are described. These include generalized classical statistical procedures as well as generalized Bayesian inference procedures.

A software system for statistical analysis of fuzzy data (AFD) is under development. Some procedures are already available, and others are in progress. The available software can be obtained from the author.

Last but not least I want to thank all persons who contributed to this work: Dr D. Hareter, Mr H. Schwarz, Mrs D. Vater, Dr I. Meliconi, H. Kay, P. Sinha-Sahay and B. Kaur from Wiley for the excellent cooperation, and my wife Dorothea for preparing the files for the last two parts of this book.

I hope the readers will enjoy the text.

Reinhard Viertl
Vienna, Austria
July 2010

Part I

FUZZY INFORMATION

Fuzzy information is a special kind of information and information is an omnipresent word in our society. But in general there is no precise definition of information.

However, in the context of statistics which is connected to uncertainty, a possible definition of information is the following: Information is everything which has influence on the assessment of uncertainty by an analyst. This uncertainty can be of different types: data uncertainty, nondeterministic quantities, model uncertainty, and uncertainty of a priori information.

Measurement results and observational data are special forms of information. Such data are frequently not precise numbers but more or less nonprecise, also called *fuzzy*. Such data will be considered in the first chapter.

Another kind of information is probabilities. Standard probability theory is considering probabilities to be numbers. Often this is not realistic, and in a more general approach probabilities are considered to be so-called *fuzzy numbers*.

The idea of generalized sets was originally published in Menger (1951) and the term 'fuzzy set' was coined in Zadeh (1965).

1

Fuzzy data

All kinds of data which cannot be presented as precise numbers or cannot be precisely classified are called nonprecise or *fuzzy*. Examples are data in the form of linguistic descriptions like high temperature, low flexibility and high blood pressure. Also, precision measurement results of continuous variables are not precise numbers but *always* more or less fuzzy.

1.1   One-dimensional fuzzy data

Measurement results of one-dimensional continuous quantities are frequently idealized to be numbers times a measurement unit. However, real measurement results of continuous quantities are never precise numbers but always connected with uncertainty. Usually this uncertainty is considered to be statistical in nature, but this is not suitable since statistical models are suitable to describe variability. For a single measurement result there is no variability, therefore another method to model the measurement uncertainty of individual measurement results is necessary. The best up-to-date mathematical model for that are so-called *fuzzy numbers* which are described in Section 2.1 [cf. Viertl (2002)].

Examples of one-dimensional fuzzy data are lifetimes of biological units, length measurements, volume measurements, height of a tree, water levels in lakes and rivers, speed measurements, mass measurements, concentrations of dangerous substances in environmental media, and so on.

A special kind of one-dimensional fuzzy data are data in the form of intervals $[a; b] \subseteq \mathbb{R}$. Such data are generated by digital measurement equipment, because they have only a finite number of digits.

Statistical Methods for Fuzzy Data   Reinhard Viertl
© 2011 John Wiley & Sons, Ltd

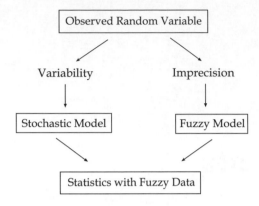

Figure 1.1   Variability and fuzziness.

1.2   Vector-valued fuzzy data

Many statistical data are multivariate, i.e. ideally the corresponding measurement results are real vectors $(x_1, \ldots, x_k) \in \mathbb{R}^k$. In applications such data are frequently not precise vectors but to some degree fuzzy. A mathematical model for this kind of data is so-called *fuzzy vectors* which are formalized in Section 2.2.

Examples of vector valued fuzzy data are locations of objects in space like positions of ships on radar screens, space–time data, multivariate nonprecise data in the form of vectors $(x_1^*, \ldots, x_n^*)$ of fuzzy numbers x_i^*.

1.3   Fuzziness and variability

In statistics frequently so-called stochastic quantities (also called random variables) are observed, where the observed results are fuzzy. In this situation two kinds of uncertainty are present: Variability, which can be modeled by probability distributions, also called stochastic models, and fuzziness, which can be modeled by fuzzy numbers and fuzzy vectors, respectively. It is important to note that these are two different kinds of uncertainty. Moreover it is necessary to describe fuzziness of data in order to obtain realistic results from statistical analysis. In Figure 1.1 the situation is graphically outlined.

Real data are also subject to a third kind of uncertainty: errors. These are the subject of Section 1.4.

1.4   Fuzziness and errors

In standard statistics errors are modeled in the following way. The observation y of a stochastic quantity is not its true value x, but superimposed by a quantity e, called error, i.e.

$$y = x + e.$$

The error is considered as the realization of another stochastic quantity. These kinds of errors are denoted as random errors.

For one-dimensional quantities, all three quantities x, y, and e are, after the experiment, real numbers. But this is not suitable for continuous variables because the observed values y are fuzzy.

It is important to note that all three kinds of uncertainty are present in real data. Therefore it is necessary to generalize the mathematical operations for real numbers to the situation of fuzzy numbers.

1.5   Problems

(a) Find examples of fuzzy numerical data which are not given in Section 1.1 and Section 1.2.
(b) Work out the difference between stochastic uncertainty and fuzziness of individual observations.
(c) Make clear how data in the form of intervals are obtained by digital measurement devices.
(d) What do X-ray pictures and data from satellite photographs have in common?

2

Fuzzy numbers and fuzzy vectors

Taking care of the fuzziness of data described in Chapter 1 it is necessary to have a mathematical model to describe such data in a quantitative way. This is the subject of Chapter 2.

2.1 Fuzzy numbers and characterizing functions

In order to model one-dimensional fuzzy data the best up-to-date mathematical model is so-called fuzzy numbers.

Definition 2.1: A fuzzy number x^* is determined by its so-called characterizing function $\xi(\cdot)$ which is a real function of one real variable x obeying the following:

(1) $\xi : \mathbb{R} \to [0; 1]$.

(2) $\forall \delta \in (0; 1]$ the so-called δ-cut $C_\delta(x^*) := \{x \in \mathbb{R} : \xi(x) \geq \delta\}$ is a finite union of compact intervals, $[a_{\delta,j}; b_{\delta,j}]$, i.e. $C_\delta(x^*) = \bigcup_{j=1}^{k_\delta} [a_{\delta,j}; b_{\delta,j}] \neq \emptyset$.

(3) The support of $\xi(\cdot)$, defined by $\mathrm{supp}[\xi(\cdot)] := \{x \in \mathbb{R} : \xi(x) > 0\}$ is bounded.

The set of all fuzzy numbers is denoted by $\mathcal{F}(\mathbb{R})$.

For the following and for applications it is important that characterizing functions can be reconstructed from the family $(C_\delta(x^*); \delta \in (0; 1])$, in the way described in Lemma 2.1.

Statistical Methods for Fuzzy Data   Reinhard Viertl
© 2011 John Wiley & Sons, Ltd

Lemma 2.1: For the characterizing function $\xi(\cdot)$ of a fuzzy number x^* the following holds true:

$$\xi(x) = \max\left\{\delta \cdot I_{C_\delta(x^*)}(x) : \delta \in [0;1]\right\} \quad \forall x \in \mathbb{R}$$

Proof: For fixed $x_0 \in \mathbb{R}$ we have

$$\delta \cdot I_{C_\delta(x^*)}(x_0) = \delta \cdot I_{\{x:\xi(x)\geq\delta\}}(x_0) = \begin{cases} \delta & \text{for} \quad \xi(x_0) \geq \delta \\ 0 & \text{for} \quad \xi(x_0) < \delta. \end{cases}$$

Therefore we have for every $\delta \in [0;1]$

$$\delta \cdot I_{C_\delta(x^*)}(x_0) \leq \xi(x_0),$$

and further

$$\sup\left\{\delta \cdot I_{C_\delta(x^*)}(x_0) : \delta \in [0;1]\right\} \leq \xi(x_0).$$

On the other hand we have for $\delta_0 = \xi(x_0)$:

$\delta_0 \cdot I_{C_{\delta_0}(x^*)}(x_0) = \delta_0$ and therefore

$\sup\left\{\delta \cdot I_{C_\delta(x^*)}(x_0) : \delta \in [0;1]\right\} \geq \delta_0$ which implies

$\sup\left\{\delta \cdot I_{C_\delta(x^*)}(x_0) : \delta \in [0;1]\right\} = \max\left\{\delta \cdot I_{C_\delta(x^*)}(x_0) : \delta \in [0;1]\right\} = \delta_0.$

Remark 2.1: In applications fuzzy numbers are represented by a finite number of δ-cuts.

Special types of fuzzy numbers are useful to define so-called fuzzy probability distribution. These kinds of fuzzy numbers are denoted as fuzzy intervals.

Definition 2.2: A fuzzy number is called a *fuzzy interval* if all its δ-cuts are non-empty closed bounded intervals.

In Figure 2.1 examples of fuzzy intervals are depicted.

The set of all fuzzy intervals is denoted by $\mathcal{F}_I(\mathbb{R})$.

Remark 2.2: Precise numbers $x_0 \in \mathbb{R}$ are represented by its characterizing function $I_{\{x_0\}}(\cdot)$, i.e. the one-point indicator function of the set $\{x_0\}$. For this characterizing function the δ-cuts are the degenerated closed interval $[x_0; x_0]. = \{x_0\}$. Therefore precise data are specialized fuzzy numbers.

In Figure 2.2 the δ-cut for a characterizing function is explained.

Special types of fuzzy intervals are so-called *LR*- fuzzy numbers which are defined by two functions $L : [0;\infty) \to [0;1]$ and $R : [0,\infty) \to [0,1]$ obeying the following:

(1) $L(\cdot)$ and $R(\cdot)$ are left-continuous.

(2) $L(\cdot)$ and $R(\cdot)$ have finite support.

(3) $L(\cdot)$ and $R(\cdot)$ are monotonic nonincreasing.

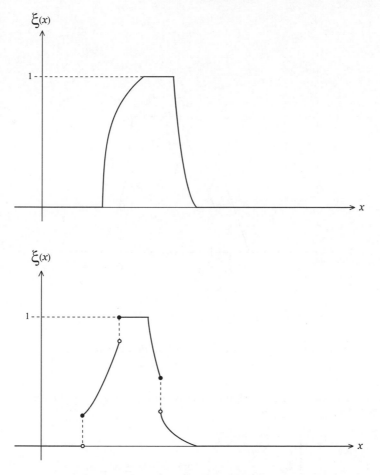

Figure 2.1   Characterizing functions of fuzzy intervals.

Using these functions the characterizing function $\xi(\cdot)$ of an *LR*-fuzzy interval is defined by:

$$\xi(x) = \begin{cases} L\left(\dfrac{m-s-x}{l}\right) & \text{for} \quad x < m-s \\ 1 & \text{for} \quad m-s \leq x \leq m+s \\ R\left(\dfrac{m-s-x}{r}\right) & \text{for} \quad x > m+s, \end{cases}$$

where m, s, l, r are real numbers obeying $s \geq 0, l > 0, r > 0$. Such fuzzy numbers are denoted by $\langle m, s, l, r \rangle_{LR}$.

A special type of *LR*-fuzzy numbers are the so-called *trapezoidal fuzzy numbers*, denoted by $t^*(m, s, l, r)$ with

$$L(x) = R(x) = \max\{0, 1-x\} \quad \forall x \in [0; \infty).$$

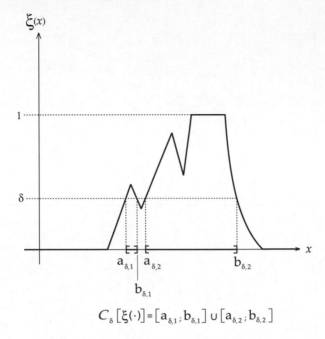

$$C_\delta\left[\xi(\cdot)\right] = \left[a_{\delta,1}; b_{\delta,1}\right] \cup \left[a_{\delta,2}; b_{\delta,2}\right]$$

Figure 2.2   Characterizing function and a δ-cut.

The corresponding characterizing function of $t^*(m, s, l, r)$ is given by

$$\xi(x) = \begin{cases} \dfrac{x - m + s + l}{l} & \text{for} \quad m - s - l \le x < m - s \\ 1 & \text{for} \quad m - s \le x \le m + s \\ \dfrac{m + s + r - x}{r} & \text{for} \quad m + s < x \le m + s + r \\ 0 & \text{otherwise.} \end{cases}$$

In Figure 2.3 the shape of a trapezoidal fuzzy number is depicted.

The δ-cuts of trapezoidal fuzzy numbers can be calculated easily using the so-called pseudo-inverse functions $L^{-1}(\cdot)$ and $R^{-1}(\cdot)$ which are given by

$$L^{-1}(\delta) = \max\left\{x \in \mathbb{R} : L(x) \ge \delta\right\}$$
$$R^{-1}(\delta) = \max\left\{x \in \mathbb{R} : R(x) \ge \delta\right\}$$

Lemma 2.2: The δ-cuts $C_\delta(x^*)$ of an *LR*-fuzzy number x^* are given by

$$C_\delta(x^*) = [m - s - l\, L^{-1}(\delta);\ m + s + r\, R^{-1}(\delta)] \quad \forall \delta \in (0, 1].$$

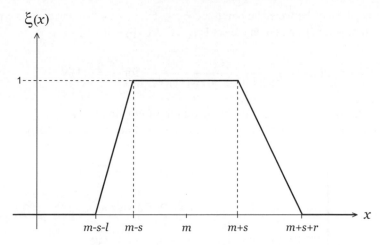

Figure 2.3   Trapezoidal fuzzy number.

Proof: The left boundary of $C_\delta(x^*)$ is determined by $\min\{x : \xi(x) \geq \delta\}$. By the definition of *LR*-fuzzy numbers for $l > 0$ we obtain

$$\min\{x : \xi(x) \geq \delta\} = \min\left\{x : L\left(\frac{m-s-x}{l}\right) \geq \delta \text{ and } x < m-s\right\}$$

$$= \min\left\{x : \frac{m-s-x}{l} \leq L^{-1}(\delta) \text{ and } x < m-s\right\}$$

$$= \min\left\{x : x \geq m-s-l\,L^{-1}(\delta) \text{ and } x < m-s\right\}$$

$$= m-s-l\,L^{-1}(\delta).$$

The proof for the right boundary is analogous.

An important topic is how to obtain the characterizing function of fuzzy data. There is no general rule for that, but for different important measurement situations procedures are available.

For analog measurement equipment often the result is obtained as a light point on a screen. In this situation the light intensity on the screen is used to obtain the characterizing function. For one-dimensional quantities the light intensity $h(\cdot)$ is normalized, i.e.

$$\xi(x) := \frac{h(x)}{\max\{h(x) : x \in \mathbb{R}\}} \quad \forall x \in \mathbb{R},$$

and $\xi(\cdot)$ is the characterizing function of the fuzzy observation.

For light points on a computer screen the function $h(\cdot)$ is given on finitely many pixels $x_1, \ldots, x_N$ with intensities $h(x_i)$, $i = 1(1)N$. In order to obtain the characterizing function $\xi(\cdot)$ we consider the discrete function $h(\cdot)$ defined on the finite set $\{x_1, \ldots, x_N\}$.

Let the distance between the points $x_1 < x_2 < \ldots < x_N$ be constant and equal to Δx. Defining a function $\eta(\cdot)$ on the set $\{x_1, \ldots, x_N\}$ by

$$\eta(x_i) := \frac{h(x_i)}{\max\{h(x_i) : i = 1(1)N\}} \quad \text{for } i = 1(1)N,$$

the characterizing function $\xi(\cdot)$ is obtained in the following way:

Based on the function $\eta(\cdot)$ the values $\xi(x)$ are defined for all $x \in \mathbb{R}$ by

$$\xi(x) := \begin{cases} 0 & \text{for} \quad x < x_1 - \dfrac{\Delta x}{2} \\[2mm] \eta(x_1) & \text{for} \quad x \in \left[x_1 - \dfrac{\Delta x}{2}; x_1 + \dfrac{\Delta x}{2}\right) \\[2mm] \max\{\eta(x_1), \eta(x_2)\} & \text{for} \quad x = x_1 + \dfrac{\Delta x}{2} \\[2mm] \vdots & \\[2mm] \max\{\eta(x_{i-1}), \eta(x_i)\} & \text{for} \quad x = x_i + \dfrac{\Delta x}{2} \\[2mm] \eta(x_i) & \text{for} \quad x \in \left(x_i - \dfrac{\Delta x}{2}; x_i + \dfrac{\Delta x}{2}\right) \\[2mm] \max\{\eta(x_i), \eta(x_{i+1})\} & \text{for} \quad x = x_i + \dfrac{\Delta x}{2} \\[2mm] \vdots & \\[2mm] \max\{\eta(x_{N-1}), \eta(x_N)\} & \text{for} \quad x = x_N - \dfrac{\Delta x}{2} \\[2mm] \eta(x_N) & \text{for} \quad x \in \left(x_N - \dfrac{\Delta x}{2}; x_N + \dfrac{\Delta x}{2}\right) \\[2mm] 0 & \text{for} \quad x > x_N + \dfrac{\Delta x}{2}. \end{cases}$$

Remark 2.3: $\xi(\cdot)$ is a characterizing function of a fuzzy number.

For digital measurement displays the results are decimal numbers with a finite number of digits. Let the resulting number be y, then the remaining (infinite many) decimals are unknown. The numerical information contained in the result is an interval $[\underline{x}; \overline{x}]$, where $\underline{x}$ is the real number obtained from y by taking the remaining decimals all to be 0, and $\overline{x}$ is the real number obtained from y by taking the remaining decimals all to be 9. The corresponding characterizing function of this fuzzy number is the indicator function $I_{[\underline{x};\overline{x}]}(\cdot)$ of the closed interval $[\underline{x}; \overline{x}]$.

Therefore the characterizing function $\xi(\cdot)$ is given by its values

$$\xi(x) := \left\{ \begin{array}{lll} 0 & \text{for} & x < \underline{x}, \\ 1 & \text{for} & \underline{x} \leq x \leq \overline{x} \\ 0 & \text{for} & x > \overline{x} \end{array} \right\} \quad \text{for all } x \in \mathbb{R}.$$

If the measurement result is given by a color intensity picture, for example diameters of particles, the color intensity is used to obtain the characterizing function $\xi(\cdot)$. Let $h(\cdot)$ be the color intensity describing the boundary then the derivative $h'(\cdot)$ of $h(\cdot)$ is used, i.e.

$$\xi(x) := \frac{|h'(x)|}{\max\{|h'(x)| : x \in \mathbb{R}\}} \quad \forall x \in \mathbb{R}.$$

An example is given in Figure 2.4.

For color intensity pictures on digital screens a discrete version for step functions can be applied.

Let $\{x_1, \ldots, x_N\}$ be the discrete values of the variable and $h(x_i)$ be the color intensities at position x_i as before, and Δx the constant distance between the points x_i. Then the discrete analog of the derivative $h'(\cdot)$ is given by the step function $\eta(\cdot)$

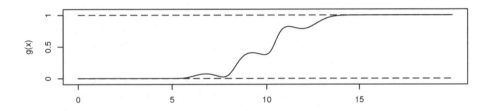

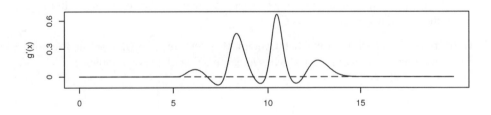

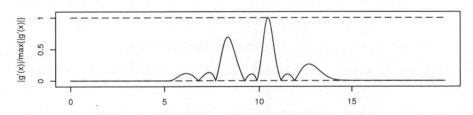

Figure 2.4   Construction of a characterizing function.

which is constant in the intervals $\left(x_i - \frac{\Delta x}{2}; x_i + \frac{\Delta x}{2}\right)$:

$$\eta(x) := |h(x_i - \varepsilon) - h(x_i + \varepsilon)| \text{ with } 0 < \varepsilon < \frac{\Delta x}{2}$$

The corresponding characterizing function $\xi(\cdot)$ of the fuzzy boundary is given by

$$\xi(x) := \frac{\eta(x)}{\max\{\eta(x) : x \in \mathbb{R}\}} \quad \text{for all } x \in \mathbb{R}.$$

The function $\xi(\cdot)$ is a characterizing function of a fuzzy number.

2.2 Vectors of fuzzy numbers and fuzzy vectors

For multivariate continuous data measurement results are fuzzy too. In the idealized case the result is a k-dimensional real vector $(x_1, \ldots, x_k)$. Depending on the problem two kinds of situations are possible.

The first is when the individual values of the variables x_i are fuzzy numbers x_i^*. Then a vector $(x_1^*, \ldots, x_k^*)$ of fuzzy numbers is obtained. This vector is determined by k characterizing functions $\xi_1(\cdot), \ldots, \xi_k(\cdot)$. Methods to obtain these characterizing functions are described in Section 2.1.

The second situation yields a fuzzy version of a vector, for example the position of a ship on a radar screen. In the idealized case the position is a two-dimensional vector $(x, y) \in \mathbb{R}^2$. In real situation the position is characterized by a light point on the screen which is not a precise vector. The result is a so-called *fuzzy vector*, denoted as $(x, y)^*$.

The mathematical model of a fuzzy vector is given in the following definition, using the notation $\underline{x} = (x_1, \ldots, x_k)$.

Definition 2.3: A k-dimensional fuzzy vector $\underline{x}^*$ is determined by its so-called vector-characterizing function $\xi(, \ldots,)$ which is a real function of k real variables $x_1, \ldots, x_k$ obeying the following:

(1) $\xi : \mathbb{R}^k \to [0; 1]$.

(2) The support of $\xi(, \ldots,)$ is a bounded set.

(3) $\forall \delta \in (0; 1]$ the so-called δ-cut $C_\delta(\underline{x}^*) := \{\underline{x} \in \mathbb{R}^k : \xi(\underline{x}) \geq \delta\}$ is non-empty, bounded, and a finite union of simply connected and closed bounded sets.

The set of all k-dimensional fuzzy vectors is denoted by $\mathcal{F}(\mathbb{R}^k)$.

In Figure 2.5 a vector-characterizing function of a two-dimensional fuzzy vector is depicted.

Remark 2.4: There are different definitions of fuzzy vectors. The above definition seems to be best for applications.

$\zeta(x_1, x_2)$

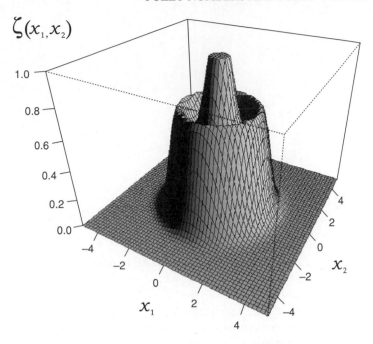

Figure 2.5   Vector-characterizing function.

For statistical inference specialized fuzzy vectors are important, the so-called fuzzy k-dimensional intervals.

Definition 2.4: A k-dimensional fuzzy vector is called a k-dimensional fuzzy interval if all δ-cuts are simply connected compact sets.

An example of a two-dimensional fuzzy interval is given in Figure 2.6.

Lemma 2.3: The vector-characterizing function $\xi(,\ldots,)$ of a fuzzy vector $\underline{x}^*$ can be reproduced by its δ-cuts in the following way:

$$\xi\left(\underline{x}\right) = \max\left\{\delta \cdot I_{C_\delta(\underline{x}^*)}\left(\underline{x}\right) \, : \, \delta \in (0; 1]\right\} \quad \forall \underline{x} \in \mathbb{R}^k$$

The proof is similar to the proof of Lemma 2.1.

Again it is important how to obtain the vector-characterizing function of a fuzzy vector. There is no general rule for this but some methodology.

For two-dimension fuzzy vectors $\underline{x}^* = (x, y)^*$ given by light intensities the vector-characterizing function $\xi(\cdot)$ of $\underline{x}^*$ is obtained from the values $h(x, y)$ of the light intensity by

$$\xi\left(x, y\right) := \frac{h\left(x, y\right)}{\max\left\{h\left(x, y\right) : (x, y) \in \mathbb{R}^2\right\}} \quad \forall\, (x, y) \in \mathbb{R}^2.$$

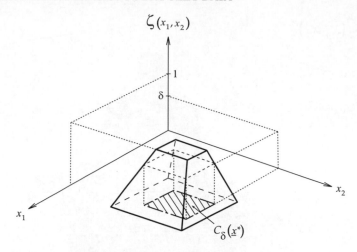

Figure 2.6   Fuzzy two-dimensional interval.

If only fuzzy values of the components x_i of a k-dimensional vector $(x_i, \dots, x_k)$ are available, i.e. x_i^* with corresponding characterizing function $\xi_i(\cdot)$, $i = 1(1)n$, these characterizing functions can be combined into a vector-characterizing function of a k-dimensional fuzzy vector using so-called *triangular norms*. Details of this are explained in Section 2.3.

2.3   Triangular norms

A vector $(x_1^*, \dots, x_k^*)$ of fuzzy numbers x_i^* is not a fuzzy vector. For the generalization of statistical inference, functions defined on sample spaces are essential. Therefore it is basic to form fuzzy elements in the sample space $M \times \dots \times M$, where M denotes the observation space of a random quantity. These fuzzy elements are fuzzy vectors. By this it is necessary to form fuzzy vectors from fuzzy samples. This is done by applying so-called triangular norms, also called t-norms.

Definition 2.5: A function $T : [0; 1] \times [0; 1] \to [0; 1]$ is called a t-norm, if for all $x, y, z, \in [0; 1]$ the following conditions are fulfilled:

(1) $T(x, y) = T(y, x)$ commutativity.

(2) $T(T(x, y), z) = T(x, T(y, z))$ associativity.

(3) $T(x, 1) = x$.

(4) $x \leq y \Rightarrow T(x, z) \leq T(y, z)$.

Examples of t-norms are:

(a) Minimum t-norm

$$T(x, y) = \min\{x, y\} \quad \forall x, y \in [0; 1]$$

(b) Product t-norm

$$T(x, y) = x \cdot y \quad \forall x, y \in [0; 1]$$

(c) Limited sum t-norm

$$T(x, y) = \max\{x + y - 1, 0\} \quad \forall x, y \in [0; 1]$$

Remark 2.5: For statistical and algebraic calculations with fuzzy data the minimum t-norm is optimal. For the combination of fuzzy components of vector data in some examples the product t-norm is more suitable.

For more details on mathematical aspects of t-norms see Klement *et al.* (2000).

Based on t-norms the combination of fuzzy numbers into a fuzzy vector is possible. For two fuzzy numbers x^* and y^* with corresponding characterizing functions $\xi(\cdot)$ and $\eta(\cdot)$ a fuzzy vector $\underline{x}^* = (x, y)^*$ is given by its vector-characterizing function $\xi(\cdot)$. The values $\xi(x, y)$ are defined based on a t-norm T by

$$\xi(x, y) := T(\xi(x), \eta(y)) \quad \forall (x, y) \in \mathbb{R}^2.$$

By the associativity of t-norms this can be extended also to k fuzzy numbers $x_1^*, \ldots, x_k^*$ with corresponding characterizing functions $\xi_1(\cdot), \ldots, \xi_k(\cdot)$ in the following way:

$$\xi(x_1, \ldots, x_k) := T(\xi_1(x_1), T(\cdots, T(\xi_{k-1}(x_{k-1}), \xi_k(x_k))\cdots)) \forall (x_1, \ldots, x_k) \in \mathbb{R}^k$$

Remark 2.6: For the minimum t-norm it follows

$$\xi(x_1, \ldots, x_k) = \min\{\xi_1(x_1), \ldots, \xi_k(x_k)\} \quad \forall (x_1, \ldots, x_k) \in \mathbb{R}^k.$$

For the minimum t-norm the δ-cuts of the combined fuzzy vector are very easy to obtain. This is shown in Proposition 2.1.

Proposition 2.1: For k fuzzy numbers $x_1^*, \ldots, x_k^*$ with characterizing functions $\xi_1(\cdot), \ldots, \xi_k(\cdot)$ the δ-cuts $C_\delta(\underline{x}^*)$ of the fuzzy vector $\underline{x}^*$, combined by the minimum t-norm are the Cartesian products of the δ-cuts $C_\delta(x_i^*)$ of the fuzzy numbers $x_1^*, \ldots, x_k^*$:

$$C_\delta(\underline{x}^*) = C_\delta(x_1^*) \times C_\delta(x_2^*) \times \cdots \times C_\delta(x_k^*) \quad \forall \delta \in (0; 1].$$

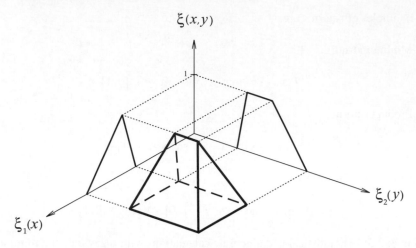

Figure 2.7   Combination of two fuzzy numbers.

Proof: Let $\xi(., \ldots ,.)$ be the vector-characterizing function of $\underline{x}^*$. Then for $\delta \in (0; 1]$ the δ-cut $C_\delta(\underline{x}^*)$ obeys

$$
\begin{aligned}
C_\delta\left(\underline{x}^*\right) &= \left\{\underline{x} \in \mathbb{R}^k : \xi\left(\underline{x}\right) \geq \delta\right\} \\
&= \left\{(x_1, \cdots, x_k) \in \mathbb{R}^k : \min\left\{\xi_1\left(x_1\right), \cdots, \xi_k\left(x_k\right)\right\} \geq \delta\right\} \\
&= \left\{(x_1, \cdots, x_k) \in \mathbb{R}^k : \xi_i\left(x_i\right) \geq \delta \quad \forall i = 1\,(1)\,k\right\} \\
&= \mathop{X}_{i=1}^{k} \left\{x_i \in \mathbb{R} : \xi_i\left(x_i\right) \geq \delta\right\} = \mathop{X}_{i=1}^{k} C_\delta\left(x_i^*\right).
\end{aligned}
$$

In Figure 2.7 this is explained graphically.

Remark 2.7: The minimum t-norm is also the natural t-norm for the generalization of algebraic operations to the system of fuzzy numbers.

2.4   Problems

(a) Draw a graph of a characterizing function and construct five δ-cuts for this characterizing function, with $\delta = 0.2, 0.4, 0.5, 0.7$ and 0.9.

(b) Let the color intensity for a one-dimensional quantity be given by the following function:

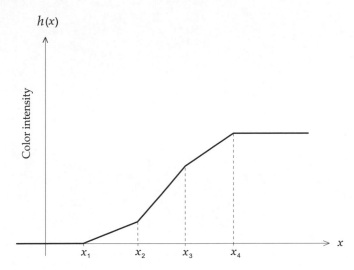

$h(x)$

Color intensity

x_1 x_2 x_3 x_4

x

Find the characterizing function of the fuzzy boundary between low color intensity and maximal color intensity using the method described in Section 2.1.

(c) Calculate the δ-cuts of a fuzzy vector with vector-characterizing function

$$\xi(x, y) = I_{[a_1;b_1] \times [a_2;b_2]}(x, y) \quad \forall (x, y) \in \mathbb{R}^2.$$

(d) Explain that the combination of fuzzy numbers, based on the minimum t-norm, yields a fuzzy vector.

(e) Prove the following: The combination of k fuzzy intervals, based on the minimum t-norm, yields a fuzzy k-dimensional interval.

3

Mathematical operations for fuzzy quantities

The extension of algebraic operations to fuzzy numbers is based on the so-called *extension principle* from fuzzy set theory. This extension principle is a method to extend classical functions $f : M \to N$ to the situation, when the argument is a fuzzy element in M.

3.1 Functions of fuzzy variables

In order to extend statistical functions $f(x_1, \ldots x_n)$ of samples $x_1, \ldots, x_n$ the following definition is useful.

Definition 3.1: (Extension principle): Let $f : M \to N$ be an arbitrary function from M to N. For a fuzzy element x^* in M, characterized by an arbitrary membership function $\mu : M \to [0; 1]$ the generalized (fuzzy) value $y^* = f(x^*)$ for the fuzzy argument value x^* is the fuzzy element y^* in N whose membership function $\psi(\cdot)$ is defined by

$$\psi(y) := \begin{cases} \sup\{\mu(x) : x \in M, f(x) = y\} & \text{if } \exists x : f(x) = y \\ 0 & \text{if } \nexists x : f(x) = y \end{cases} \quad \forall y \in N.$$

Remark 3.1: The extension principle is a natural way to model the propagation of imprecision. It corresponds to the engineering principle 'to be on the safe side'.

The following statements are important for the generalization of statistical functions.

Proposition 3.1: Let $f : M \to N$ be a classical function, and x^* a fuzzy element of M with membership function $\mu(\cdot)$. Then the fuzzy element $y^* = f(x^*)$ defined by the extension principle obeys the following:

$$f\left(C_\delta(x^*)\right) \subseteq C_\delta(y^*) \quad \forall \delta \in (0; 1]$$

Proof: $C_\delta(y^*) = \{y \in N : \psi(y) \geq \delta\}$ where $\psi(\cdot)$ is defined as in Definition 3.1. By $f\left(C_\delta(x^*)\right) = \{f(x) : x \in C_\delta(x^*)\}$, for $y \in (C_\delta(x^*))$ there exists $x \in (C_\delta(x^*))$ with $f(x) = y$. Therefore $\mu(x) \geq \delta$ and $\sup\{\mu(x) : f(x) = y\} \geq \delta$ and $y \in (C_\delta(y^*))$.

Classical statistical functions are frequently functions $f : \mathbb{R}^n \to \mathbb{R}$. If such statistics are continuous functions, the following theorem holds.

Theorem 3.1: Let $f : \mathbb{R}^n \to \mathbb{R}$ be a continuous function and $\underline{x}^*$ a fuzzy n-dimensional interval. Then the following is true:

(a) $f(\underline{x}^*)$ defined by the extension principle is a fuzzy interval.

(b) $C_\delta\left[f(\underline{x}^*)\right] = \left[\min_{\underline{x} \in C_\delta(\underline{x}^*)} f(\underline{x}); \max_{\underline{x} \in C_\delta(\underline{x}^*)} f(\underline{x})\right]$.

Proof: Let $\xi(\cdot)$ be the vector-characterizing function of $\underline{x}^*$. By the continuity of $f(\cdot)$, for all $y \in \mathbb{R}$ the set $f^{-1}(\{y\})$ is closed. Next we prove $C_\delta[f(\underline{x}^*)] = f(C_\delta(\underline{x}^*)) \; \forall \delta \in (0; 1]$. By Proposition 3.1 we have

$$f\left(C_\delta(\underline{x}^*)\right) \subseteq C_\delta\left[f(\underline{x}^*)\right] \quad \forall \delta \in (0; 1].$$

In order to prove the converse inclusion let $y \in C_\delta[f(\underline{x}^*)]$. By the extension principle we have $\sup\{\xi(\underline{x}) : \underline{x} \in f^{-1}(\{y\})\} \geq \delta$. Therefore there exists a sequence $\underline{x}_n \in f^{-1}(\{y\}), n \in \mathbb{N}$ with $\sup\{\xi(\underline{x}_n) : n \in \mathbb{N}\} \geq \delta$. ($\mathbb{N}$ is the set of all natural numbers, i.e. $\mathbb{N} = \{1, 2, 3, \ldots\}$).

Since $f^{-1}(\{y\})$ is a closed subset of $\mathbb{R}^n$, $f^{-1}(\{y\}) \cap C_\delta(\underline{x}^*)$ is compact. If there exists an $\underline{x}_n$ in the sequence with $\xi(\underline{x}_n \geq \delta)$ then $\psi(y) \geq \delta$ and therefore $y \in C_\delta(f(\underline{x}^*))$. If $\xi(\underline{x}_n) < \delta$ for all $\underline{x}_n, n \in \mathbb{N}$, then by $\sup\{\xi(\underline{x}_n) : n \in \mathbb{N}\} \geq \delta$ the sequence $\xi(\underline{x}_n)$ has accumulation point δ. Therefore there exists a convergent subsequence $z_n, n \in \mathbb{N}$, and by the compactness of $f^{-1}(\{y\}) \cap C_\delta(\underline{x}^*)$ the limit z_0 of this subsequence belongs to $f^{-1}(\{y\}) \cap C_\delta(\underline{x}^*)$. By $\lim_{n \to \infty} \xi(z_n) = \delta$ we obtain $\xi(z_0) = \delta$ and therefore $\psi(y) = \delta$, and $y \in f(C_\delta(\underline{x}^*))$. By definition of fuzzy vectors $C_\delta(\underline{x}^*)$ is compact and connected. By the continuity of $f(\cdot)$ follows that $f(C_\delta(\underline{x}^*))$ is connected and compact. Therefore it is a closed finite interval.

Corollary 3.1: For continuous function $f : \mathbb{R} \to \mathbb{R}$ and fuzzy interval x^*, the fuzzy value $f(x^*)$ is a fuzzy interval with

$$C_\delta\left[f(x^*)\right] = \left[\min_{x \in C_\delta(x^*)} f(x); \max_{x \in C_\delta(x^*)} f(x)\right] \quad \forall \delta \in (0; 1].$$

In Figure 3.1 an example of Corollary 3.1 is depicted.

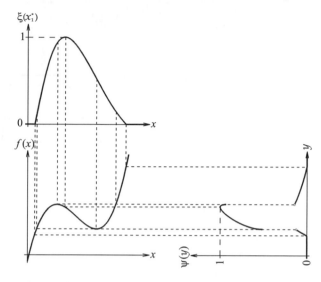

Figure 3.1   Continuous function of a fuzzy interval.

Remark 3.2: The resulting characterizing function in Figure 3.1 is not continuous, although both $\xi(\cdot)$ and $f(\cdot)$ are continuous.

3.2   Addition of fuzzy numbers

Let x_1^* and x_2^* be two fuzzy numbers with corresponding characterizing functions $\xi_1(\cdot)$ and $\xi_2(\cdot)$. The generalized addition operation $\oplus$ for fuzzy numbers has to obey two demands: First it has to generalize the addition of real numbers, and secondly it has to generalize interval arithmetic.

For fuzzy intervals x_1^* and x_2^* the generalized addition can be defined using δ-cuts.

Let $C_\delta(x_1^*) = [a_{\delta,1}; b_{\delta,1}]$ and $C_\delta(x_2^*) = [a_{\delta,2}; b_{\delta,2}]$ $\forall \delta \in [0; 1]$ then the δ-cut of the fuzzy sum $x_1^* \oplus x_2^*$ is given by

$$C_\delta(x_1^* \oplus x_2^*) = [a_{\delta,1} + a_{\delta,2}; b_{\delta,1} + b_{\delta,2}]   \quad \forall \delta \in (0; 1].$$

The characterizing function of $x_1^* \oplus x_2^*$ is obtained by Lemma 2.1.

Remark 3.3: The same result is obtained if the generalized sum is defined via the extension principle, applying the function $+$, defined on the Cartesian product $\mathbb{R} \times \mathbb{R}$, where x_1^* and x_2^* are combined into a two-dimensional fuzzy interval by the minimum t-norm. This means

$$\xi_{x_1^* \oplus x_2^*}(x) = \sup \{\min\{\xi_1(x_1), \xi_2(x_2)\} :   x_1 + x_2 = x\}$$
$$= \sup \{\min\{\xi_1(y), \xi_2(x - y)\} :   y \in \mathbb{R}\}   \quad \forall x \in \mathbb{R}.$$

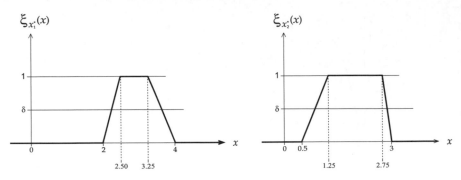

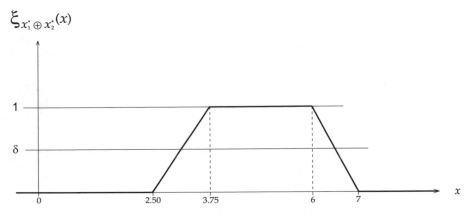

Figure 3.2   Addition of fuzzy intervals.

An example is given in Figure 3.2.

Remark 3.4: The sum of two fuzzy intervals is again a fuzzy interval.

A special case of the addition of fuzzy numbers is the translation by a constant.

For fuzzy numbers x^* with characterizing function $\xi(\cdot)$, and constant $c \in \mathbb{R}$, the *translation* $x^* \oplus c$ is the fuzzy number whose characterizing function $\eta(\cdot)$ is defined by

$$\eta(x) := \xi(x - c) \quad \forall x \in \mathbb{R}.$$

For so-called *LR*-fuzzy numbers, from Chapter 2, i.e. $x^* = \langle m, s, l, r \rangle_{LR}$, the following holds concerning the addition:

Proposition 3.2: The sum of two *LR*-fuzzy numbers $x_1^* = \langle m_1, s_1, l_1, r_1 \rangle$ and $x_2^* = \langle m_2, s_2, l_2, r_2 \rangle$ is an *LR*-fuzzy number, with

$$x_1^* \oplus x_2^* = \langle m_1 + m_2, s_1 + s_2, l_1 + l_2, r_1 + r_2 \rangle_{LR}$$

Proof: This can be proved using δ-cuts and by application of Lemma 2.2.

Corollary 3.2: The sum of trapezoidal fuzzy numbers is a trapezoidal fuzzy number.

3.3 Multiplication of fuzzy numbers

The generalized product $x_1^* \odot x_2^*$ of two fuzzy numbers with corresponding characterizing functions $\xi_1(\cdot)$ and $\xi_2(\cdot)$ is defined by using the extension principle to the function $f : \mathbb{R}^2 \rightarrow \mathbb{R}$ with $f(x, y) = x \cdot y$, after combining x_1^* and x_2^* by the minimum t-norm.

The characterizing function $\xi_{x_1^* \odot x_2^*}(\cdot)$ of the *fuzzy product* $x_1^* \odot x_1^*$ is given by

$$\xi_{x_1^* \odot x_2^*}(x) = \sup \{\min \{\xi_1(x_1), \xi_2(x_2)\} : x_1 \cdot x_2 = x\} \quad \forall x \in \mathbb{R}.$$

Remark 3.5: Using Theorem 3.1 the δ-cuts of the product $x_1^* \odot x_2^*$ can be calculated:

$$C_\delta(x_1^* \odot x_2^*) = \left[\min_{(x_1,x_2) \in C_\delta(x_1^*) \times C_\delta(x_2^*)} x_1 \cdot x_2 ; \max_{(x_1,x_2) \in C_\delta(x_1^*) \times C_\delta(x_2^*)} x_1 \cdot x_2 \right] \forall \delta \in (0; 1].$$

It can be proved that the following is valid:

Let $C_\delta(x_i^*) = [a_{\delta,i}; b_{\delta,i}]$ for $i = 1$ or 2, then
$$C_\delta(x_1^* \odot x_2^*) = [\min \{a_{\delta,1} \cdot a_{\delta,2}, a_{\delta,1} \cdot b_{\delta,2}, b_{\delta,1} \cdot a_{\delta,2}, b_{\delta,1} \cdot b_{\delta,2}\} ;$$
$$\max \{a_{\delta,1} \cdot a_{\delta,2}, a_{\delta,1} \cdot b_{\delta,2}, b_{\delta,1} \cdot a_{\delta,2}, b_{\delta,1} \cdot b_{\delta,2}\}] \quad \forall \delta \in (0; 1].$$

It should be noted that the product of trapezoidal fuzzy numbers is not trapezoidal.

In Figure 3.3 an example for the product of fuzzy numbers is given by using characterizing functions.

The special case of multiplication of a fuzzy number by a real constant $c \in \mathbb{R}$ is called *scalar multiplication* of fuzzy numbers.

For fuzzy numbers x^* with characterizing function $\xi(\cdot)$ and $c \in \mathbb{R} \setminus \{0\}$, the multiplication $c \cdot x^*$ yields a fuzzy number whose characterizing function $\eta(\cdot)$ is given by

$$\eta(x) := \xi \left(\frac{y}{c} \right) \quad \forall x \in \mathbb{R}.$$

For $c = 0$ the result $0 \cdot x^*$ is the precise number 0, i.e.

$$\eta(\cdot) = I_{\{0\}}(\cdot).$$

3.4 Mean value of fuzzy numbers

Arithmetic operations for fuzzy numbers can be extended to more than two fuzzy numbers by the associativity of *t*-norms.

This applies especially to sums of more than two fuzzy numbers.

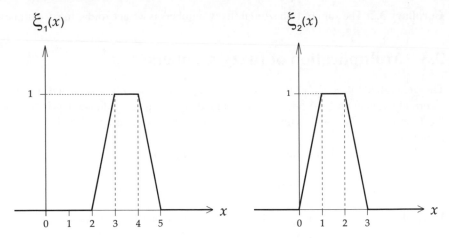

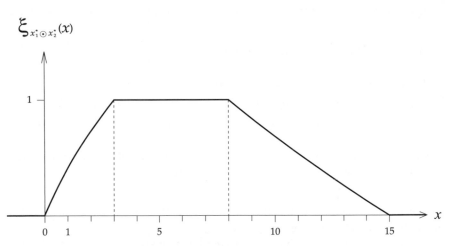

Figure 3.3   Product of two fuzzy intervals.

The sum of n fuzzy numbers $x_1^*, \ldots, x_n^*$ is denoted by

$$\bigoplus_{i=1}^{n} x_i^* = x_1^* \oplus x_2^* \oplus \cdots \oplus x_n^*.$$

Combining this with scalar multiplication yields the *arithmetic mean* of fuzzy numbers:

$$\frac{1}{n} \cdot \bigoplus_{i=1}^{n} x_i^*$$

For the arithmetic mean of n fuzzy intervals the following proposition holds.

Proposition 3.3: Let $x_1^*, \ldots, x_n^*$ be fuzzy intervals with δ-cuts $C_\delta(x_i^*) = [a_{\delta,i}; b_{\delta,i}]$, then the fuzzy arithmetic mean

$$\overline{x}_n^* = \frac{1}{n} \cdot \bigoplus_{i=1}^{n} x_i^* \text{ is a fuzzy interval, and}$$

$$C_\delta(\overline{x}_n^*) = \left[\frac{1}{n} \sum_{i=1}^{n} a_{\delta,i}; \frac{1}{n} \sum_{i=1}^{n} b_{\delta,i} \right] \quad \forall \delta \in (0, 1].$$

Proof: This is a consequence of Theorem 3.1.

Remark 3.6: The minimum t-norm is the only combination rule for which $\frac{1}{n} \cdot \bigoplus_{i=1}^{n} x^* = x^*$.

3.5   Differences and quotients

The difference $x_1^* \ominus x_2^*$ of two fuzzy numbers x_1^* and x_2^* are defined, using $-x_2^* := (-1) \cdot x_2^*$, by $x_1^* \ominus x_2^* := x_1^* \oplus (-x_2^*)$.

The quotient $x_1^* : x_2^*$ is defined by the function $f(x, y) = x/y$, provided that $0 \notin$ supp (x_2^*).

3.6   Fuzzy valued functions

Fuzzy valued functions $f^*(\cdot)$ are functions whose range of values are fuzzy intervals, i.e.

$$f^* : M \to \mathcal{F}_I(\mathbb{R}).$$

Definition 3.2: Let $f^*(\cdot)$ be a fuzzy valued function with domain M, and $x \in M$. Denoting the δ-cuts of $f^*(x)$ by

$$C_\delta \left[f^*(x) \right] = \left[\underline{f}_\delta(x); \overline{f}_\delta(x) \right] \quad \forall \delta \in (0; 1],$$

for variable x the real valued functions $\underline{f}_\delta(\cdot)$ and $\overline{f}_\delta(\cdot)$ are called δ-level functions.

In the case $M = \mathbb{R}$ the δ-level functions are called δ-level curves.

Fuzzy valued functions can be graphically displayed by depicting some δ-level curves. An example is given in Figure 3.4.

Remark 3.7: Fuzzy valued functions are important in environmetrics when modeling emissions to the environment. For calculations concerning total amounts of emissions integration of emission rates are necessary. Therefore a calculus of generalized integration is needed. This is possible based on δ-level functions with existing integrals.

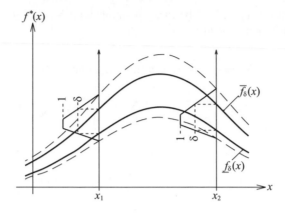

Figure 3.4   δ-Level curves of a fuzzy valued function.

Definition 3.3: Let $f^*(\cdot)$ be a fuzzy valued function defined on a measure space $(M, \mathcal{A}, \mu)$. If all δ-level functions $\underline{f}_\delta(\cdot)$ and $\overline{f}_\delta(\cdot)$ are integrable with finite integrals $\int \underline{f}_\delta(x)d\mu(x)$ and $\int \overline{f}_\delta(x)d\mu(x)$, then the generalized integral $\int f^*(x)d\mu(x)$ is the fuzzy number J^* whose δ-cuts $C_\delta(J^*)$ are defined by

$$C_\delta(J^*) = \left[\int_M \underline{f}_\delta(x)d\mu(x); \int_M \overline{f}_\delta(x)d\mu(x) \right] \quad \forall \delta \in (0; 1].$$

By the representation Lemma 2.1 the characterizing function of the fuzzy integral J^* is given by

$$\xi_{J^*}(x) = \sup \left\{ \delta \cdot I_{[\underline{J}_\delta; \overline{J}_\delta]}(x) : \delta \in [0; 1] \right\} \quad \forall x \in \mathbb{R},$$

where $\underline{J}_\delta := \int_M \underline{f}_\delta(x)d\mu(x)$ and $\overline{J}_\delta := \int_M \overline{f}_\delta(x)d\mu(x)$. The resulting fuzzy integral is denoted by $\oint f^*(x)d\mu(x)$.

Remark 3.8: For fuzzy valued real functions, i.e. $f^* : [a, b] \to \mathcal{F}_1(\mathbb{R})$ and integrable δ-level curves the *fuzzy integral* is defined by $\underline{J}_\delta = \int_a^b \underline{f}_\delta(x)dx$ and $\overline{J}_\delta := \int_a^b \overline{f}_\delta(x)dx$. The resulting fuzzy value is denoted by $\oint_a^b f^*(x)dx$.

3.7   Problems

(a) Calculate the δ-cuts of the fuzzy vector $\underline{x}^*$ which is obtained by applying the minimum t-norm to the two trapezoidal fuzzy numbers $x_1^* = t^*(2, 0, 1, 1)$ and $x_2^* = t^*(3, 1, 1, 1)$.

(b) Prove the formula for the δ-cuts of $x_1^* \odot x_2^*$ in Remark 3.3 if x_1^* and x_2^* are fuzzy intervals.

(c) Two fuzzy intervals x_1^* and x_2^* are given by their characterizing functions $\xi_1(\cdot)$ and $\xi_2(\cdot)$:

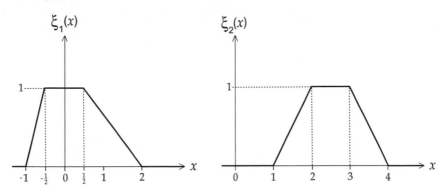

Calculate and draw the characterizing function of $x_1^* \oplus x_2^*$.

(d) Calculate and draw the characterizing function of the product $x_1^* \odot x_2^*$ for x_1^* and x_2^* in (c).

(e) Explain that the generalized integration of fuzzy valued functions reduces to the classical integration for classical real valued functions.

(f) Calculate the fuzzy integral for a fuzzy constant function $f^*(x) \equiv y^*$.

Part II

DESCRIPTIVE STATISTICS FOR FUZZY DATA

In standard statistics – objectivist and Bayesian analysis – observations are assumed to be numbers, vectors, or classical functions. By the fuzziness of real data of continuous quantities it is necessary to adapt descriptive statistical methods to the situation of fuzzy data. This is done for the most frequently used descriptive methods in this part.

4

Fuzzy samples

Fuzzy samples consist of a finite sequence of fuzzy numbers, i.e. $x_1^*, \ldots x_n^*$, or a finite sequence of fuzzy vectors $\underline{x}_1^*, \ldots \underline{x}_n^*$. In this chapter the concepts of minimum and maximum of observations are extended to fuzzy samples. Moreover generalized cumulative sums are introduced.

4.1 Minimum of fuzzy data

Let $x_1^*, \ldots x_n^*$ be n fuzzy intervals whose corresponding δ-cuts are denoted by $C_\delta(x_i^*) = [\underline{x}_{\delta,i}, \overline{x}_{\delta,i}]$ for $\delta \in (0; 1]$ and $i = 1(1)n$. The fuzzy valued minimum $\min\{x_1^*, \ldots x_n^*\}$ is a fuzzy interval $x_{\min}^*$ whose δ-cuts $C_\delta(x_{\min}^*)$ are defined by

$$C_\delta(x_{\min}^*) := \left[\min\left\{\underline{x}_{\delta,i} : i = 1(1)n\right\}; \quad \min\left\{\overline{x}_{\delta,i} : i = 1(1)n\right\}\right] \quad \forall \delta \in (0; 1].$$

The characterizing function of $x_{\min}^*$ is obtained by (the representation) Lemma 2.1.

4.2 Maximum of fuzzy data

Under the same conditions as in Section 4.1 the maximum of n fuzzy intervals $x_1^*, \ldots x_n^*$ is the fuzzy interval $x_{\max}^* = \max\{x_1^*, \ldots x_n^*\}$ whose δ-cuts $C_\delta(x_{\max}^*)$ are defined by

$$C_\delta(x_{\max}^*) = \left[\max\left\{\underline{x}_{\delta,i} : i = 1(1)n\right\}; \max\left\{\overline{x}_{\delta,i} : i = 1(1)n\right\}\right] \quad \forall \delta \in (0; 1].$$

Statistical Methods for Fuzzy Data   Reinhard Viertl
© 2011 John Wiley & Sons, Ltd

Remark 4.1: The minimum as well as maximum of fuzzy intervals reduce to the classical minimum and maximum for classical samples $x_1, \ldots, x_n$.

Examples are given in Section 4.4.

4.3   Cumulative sum for fuzzy data

In the case of classical samples $x_1, \ldots, x_n$ with $x_i \in \mathbb{R}$ of one-dimensional quantities, cumulative sums are defined by

$$\#\{x_i : x_i \le x\} \quad \forall x \in \mathbb{R},$$

and normalized cumulative sums $S_n(\cdot)$ are defined by

$$S_n(x) := \frac{\#\{x_i : x_i \le x\}}{n} \quad \forall x \in \mathbb{R}.$$

Such cumulative sums give the proportion of observations which are not larger than a given real value x.

Remark 4.2: For (idealized) precise observations $x_i \in \mathbb{R}$ the cumulative sum coincides with the *empirical distribution function* $\hat{F}_n(\cdot)$, defined by its values

$$\hat{F}_n(x) := \frac{1}{n} \sum_{i=1}^{n} I_{(-\infty;x]}(x_i) \quad \forall x \in \mathbb{R}.$$

For fuzzy observations given as fuzzy numbers x_i^*, $i = 1(1)n$, the definition of cumulative sums has to be adapted. This *adapted cumulative sum* $S_n^{ad}(\cdot)$ is defined using the characterizing functions $\xi_i(\cdot)$ of x_i^*, $i = 1(1)n$, which are assumed to be integrable, by

$$S_n^{ad}(x) := \frac{\sum\limits_{i=1}^{n} \int\limits_{-\infty}^{x} \xi_i(t)\, dt}{\sum\limits_{i=1}^{n} \int\limits_{-\infty}^{\infty} \xi_i(t)\, dt} \quad \forall x \in \mathbb{R}.$$

Remark 4.3: $S_n^{ad}(x)$ is the proportion of fuzzy observations less than or equal to x.

4.4   Problems

(a) A fuzzy sample is given by its characterizing functions $\xi_i(\cdot)$, $i = 1(1)8$, in the following figure:

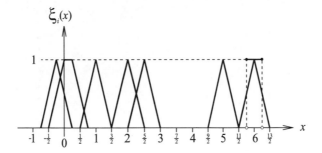

Calculate and draw the resulting fuzzy minimum and fuzzy maximum of this fuzzy sample.

(b) Make a diagram for the adapted cumulative sum function $S_n^{ad}(\cdot)$ for the fuzzy sample from (a).

5

Histograms for fuzzy data

Classical histograms are based on precise data $x_1, \ldots, x_n$ in order to explain the distribution of the observations x_i. In order to construct histograms the set M of possible values x_i is decomposed into so-called classes $K_1, \ldots, K_k$ with $K_i \cap K_j = \emptyset \quad \forall i \neq j$, and

$$\bigcup_{j=1}^{k} K_j = M.$$

Then for each class K_j the so-called *relative frequency* $h_n(K_j)$ is calculated, i.e.

$$h_n(K_j) := \frac{\#\{x_i : x_i \in K_j\}}{n}, \quad j = 1(1)k.$$

The display of the relative frequencies is called a histogram.

For fuzzy data the following problem arises: By the fuzziness of observations it is not always possible to decide in which class K_j a fuzzy observation x^* with characterizing function $\xi(\cdot)$ lies. This is depicted in Figure 5.1.

Therefore a generalization of histograms is necessary.

5.1 Fuzzy frequency of a fixed class

Due to the possible inability to decide in which class a fuzzy data point x_i^* lies it seems natural to consider fuzzy values of relative frequencies. For n fuzzy observations $x_1^*, \ldots, x_n^*$ and classes $K_1, \ldots, K_k$ the fuzzy relative frequencies of the classes are

Statistical Methods for Fuzzy Data   Reinhard Viertl
© 2011 John Wiley & Sons, Ltd

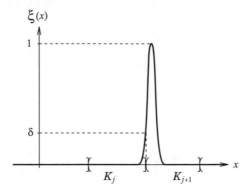

Figure 5.1   Fuzzy observation and classes of a histogram.

denoted by $h_n^*(K_j)$, $j = 1(1)k$. The characterizing function of $h_n^*(K_j)$ is constructed via its δ-cuts

$$C_\delta \left(h_n^* \left(K_j \right) \right) = \left[\underline{h}_{n,\delta}(K_j); \overline{h}_{n,\delta}(K_j) \right],$$

where the limit points $\underline{h}_{n,\delta}(K_j)$ and $\overline{h}_{n,\delta}(K_j)$ are defined in the following way:

$$\overline{h}_{n,\delta}(K_j) = \frac{\# \left\{ x_i^* : C_\delta(x_i^*) \cap K_j \neq \emptyset \right\}}{n}$$

and

$$\underline{h}_{n,\delta}(K_j) = \frac{\# \left\{ x_i^* : C_\delta(x_i^*) \subseteq K_j \right\}}{n}$$

for all $\delta \in [0; 1]$.

The characterizing function $\eta_j(\cdot)$ of $h_n^*(K_j)$ is obtained by Lemma 2.1.

Remark 5.1: The characterizing function of a fuzzy relative frequency is a step-function. In Figure 5.2 the characterizing functions of a sample of size 8 are given. In Figure 5.3 the characterizing function of the fuzzy relative frequency of the class [1; 2] is depicted.

5.2   Fuzzy frequency distributions

Considering all classes $K_1, \ldots, K_k$ of a decomposition of M, the fuzzy relative frequencies $h_n^*(K_j)$, $j = 1(1)k$ are defining a special fuzzy valued function $f^* : M \to \mathcal{F}_I(\mathbb{R})$ from Section 3.6. For every $x \in M$ the fuzzy value $f^*(x)$ is defined in the following way:

Determine the class K_j with $x \in K_j$, and define the fuzzy value

$$f^*(x) := h_n^*(K_j).$$

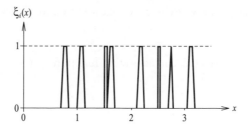

Figure 5.2   Fuzzy sample.

Remark 5.2: The abstract mathematical model for idealized fuzzy histograms are so-called *fuzzy probability distributions*, especially fuzzy probability densities which are described in Part III, and especially in Chapter 9.

Graphically fuzzy histograms can be depicted using δ-level functions, and δ-level curves for one-dimensional distributions.

An alternative way of representing fuzzy histograms is given in the next section.

Remark 5.3: Different from histograms for precise data, where relative frequencies $h_n(\cdot)$ are additive, i.e. for two disjoint sets A and B we have

$$h_n(A \cup B) = h_n(A) + h_n(B),$$

for fuzzy relative frequencies $h_n^*(K_j)$ less restrictive relations for the boundaries $\underline{h}_{n.\delta}(K_j)$ and $\overline{h}_{n,\delta}(K_j)$ of the δ-cuts are valid, i.e.

$$\overline{h}_{n,\delta}\left(K_i \cup K_j\right) \leq \overline{h}_{n,\delta}(K_i) + \overline{h}_{n,\delta}(K_j) \tag{5.1}$$

and

$$\underline{h}_{n,\delta}\left(K_i \cup K_j\right) \geq \underline{h}_{n,\delta}(K_i) + \underline{h}_{n,\delta}(K_j) \tag{5.2}$$

for all $\delta \in (0; 1]$.

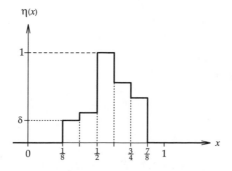

Figure 5.3   Characterizing function of $h_8^*([1; 2])$.

Moreover the fuzzy relative frequencies of the extreme events $\emptyset$ and M are precise numbers, i.e.

$$h_n^*(\emptyset) \text{ has characterizing function } I_{\{0\}}(\cdot)$$

and

$$h_n^*(M) \text{ has characterizing function } I_{\{1\}}(\cdot).$$

5.3   Axonometric diagram of the fuzzy histogram

For one-dimensional data the characterizing functions of the fuzzy relative frequencies are constant on the classes K_j. Therefore it is possible to depict them in an axonometric way. One axis is the axis on which the classes are defined. The second axis is the independent variable y for the characterizing function $\eta_j(\cdot)$ of the fuzzy relative frequency $h_n^*(K_j)$. The third axis is the axis for the values $\eta_j(y)$.

An example of the axonometric picture of a fuzzy histogram is given in Figure 5.4.

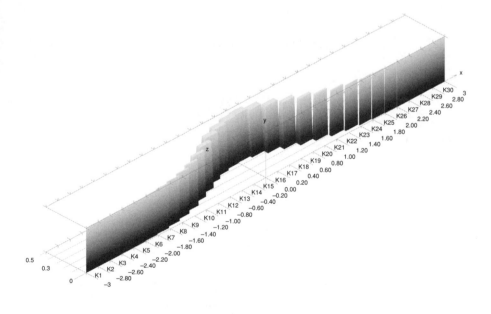

Figure 5.4   Fuzzy histogram.

5.4   Problems

(a) For the fuzzy sample with characterizing functions given by the following diagram

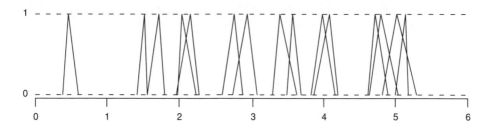

use the classes $K_1 = [0; 1]$, $K_2 = (1; 2]$, $K_3 = (2; 3]$, ..., $K_{10} = (9; 10]$ and calculate the fuzzy relative frequencies $h_{15}^*(K_j)$ for $j = 1(1)10$.

(b) Prove the inequalities (5.1) and (5.2) from Section 5.2.

6

Empirical distribution functions

The classical empirical distribution function for n precise data points $x_1, \ldots, x_n$ is defined by

$$\hat{F}_n(x) := \frac{1}{n} \sum_{i=1}^{n} I_{(-\infty;x]}(x_i) \quad \text{for all } x \in \mathbb{R}.$$

6.1 Fuzzy valued empirical distribution function

For fuzzy samples $x_1^*, \ldots, x_n^*$ with corresponding characterizing functions $\xi_1(\cdot), \ldots, \xi_n(\cdot)$ the counterpart to the empirical distribution function is a fuzzy valued function $\hat{F}_n^*(\cdot)$ defined on $\mathbb{R}$ whose lower and upper δ-level functions $\hat{F}_{\delta,L}(\cdot)$ and $\hat{F}_{\delta,U}(\cdot)$ are defined via the δ-cuts $C_\delta(x_i^*)$ of the fuzzy observations x_i^*, which are assumed to be fuzzy intervals.

For fixed $x \in \mathbb{R}$ the values $\hat{F}_{\delta,L}(x)$ and $\hat{F}_{\delta,U}(x)$ are defined by

$$\hat{F}_{\delta,U}(x) := \frac{\# \left\{ x_i^* : C_\delta(x_i^*) \cap (-\infty; x] \neq \emptyset \right\}}{n}$$

$$\hat{F}_{\delta,L}(x) := \frac{\# \left\{ x_i^* : C_\delta(x_i^*) \subseteq (-\infty; x] \right\}}{n}$$

for all $\delta \in [0; 1]$.

Statistical Methods for Fuzzy Data   Reinhard Viertl
© 2011 John Wiley & Sons, Ltd

Lemma 6.1: For fuzzy intervals $x_1^*, \ldots, x_n^*$ whose δ-cuts are given by

$$C_\delta(x_i^*) = \left[\underline{x}_{\delta,i}; \overline{x}_{\delta,i}\right] \quad \forall \delta \in (0; 1]$$

the corresponding δ-level functions of the fuzzy valued empirical distribution function are given by

$$\hat{F}_{\delta,U}(x) = \frac{1}{n} \sum_{i=1}^{n} I_{(-\infty;x]}(\underline{x}_{\delta,i})$$

and

$$\hat{F}_{\delta,L}(x) = \frac{1}{n} \sum_{i=1}^{n} I_{(-\infty;x]}(\overline{x}_{\delta,i}),$$

respectively.

Proof: For the value $\hat{F}_{\delta,U}(x)$ of the upper δ-level function $\hat{F}_{\delta,U}(\cdot)$ we have $C_\delta(x_i^*) \cap (-\infty; x] \neq \emptyset$ if and only if $\underline{x}_{\delta,i} \leq x$, and this is equivalent to $I_{(-\infty;x]}(\underline{x}_{\delta,i}) = 1$.

Analogously for $\hat{F}_{\delta,L}(x)$ we have $C_\delta(x_i^*) \subseteq (-\infty; x]$ if and only if $\overline{x}_{\delta,i} \leq x$, which is equivalent to $I_{(-\infty;x]}(\overline{x}_{\delta,i}) = 1$.

In Figure 6.1 a fuzzy sample and the lower and upper δ-level curves of the corresponding fuzzy empirical distribution function are depicted.

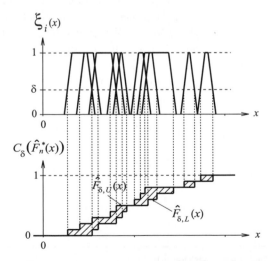

Figure 6.1   Fuzzy sample and corresponding δ-level curves of $\hat{F}_n^*(\cdot)$.

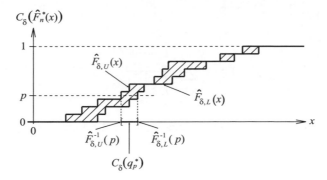

Figure 6.2   Construction of $C_\delta(q_p^*)$.

6.2   Fuzzy empirical fractiles

For classical empirical distribution functions $\hat{F}_n(\cdot)$ the empirical fractiles are defined in the following way: Let $p \in (0; 1)$ then the p-fractile x_p is given by

$$\min \left\{ x \in \mathbb{R} : \hat{F}_n(x) \geq p \right\}.$$

In the case of fuzzy empirical distribution functions $\hat{F}_n^*(\cdot)$ generalized empirical fractiles are fuzzy numbers. The characterizing function or the δ-cuts of the fuzzy empirical fractiles are defined in the following way:

For $p \in (0, 1)$ the lower and upper δ-level curves $\hat{F}_{\delta,L}(\cdot)$ and $\hat{F}_{\delta,U}(\cdot)$ are used to define the δ-cuts of the corresponding fuzzy fractile q_p^*. The δ-cut $C_\delta(q_p^*)$ is defined by

$$C_\delta(q_p^*) = \left[\hat{F}_{\delta,U}^{-1}(p); \hat{F}_{\delta,L}^{-1}(p) \right] \quad \forall \delta \in (0; 1).$$

The method is explained in Figure 6.2.

Remark 6.1: The characterizing function $\xi_{q_p^*}(\cdot)$ of the fuzzy empirical quantile q_p^* is obtained by Lemma 2.1.

6.3   Smoothed empirical distribution function

Another possibility to adapt the concept of the empirical distribution function to the situation of fuzzy data is the so-called *smoothed empirical distribution function* $\hat{F}_n^{sm}(\cdot)$.

The values $\hat{F}_n^{sm}(x)$ of the smoothed empirical distribution function based on fuzzy data $x_1^*, \ldots, x_n^*$ with characterizing functions $\xi_1(\cdot), \ldots, \xi_n(\cdot)$ are defined by generalizing the classical empirical distribution function $\hat{F}_n(\cdot)$ from Section 4.3

(cf. Remark 4.2).

$$\hat{F}_n^{sm}(x) := \frac{1}{n} \sum_{i=1}^{n} \frac{\int\limits_{-\infty}^{x} \xi_i(t)dt}{\int\limits_{-\infty}^{\infty} \xi_i(t)dt} \quad \forall x \in \mathbb{R}$$

provided that all characterizing functions $\xi_1(\cdot), \ldots, \xi_n(\cdot)$ are integrable with finite integral

$$\int\limits_{-\infty}^{\infty} \xi_i(t)dt < \infty.$$

In case the sample contains some precise values, say $x_1, \ldots, x_l$ and the rest are fuzzy values $x_{l+1}^*, \ldots, x_n^*$ with characterizing functions $\xi_{l+1}(\cdot), \ldots, \xi_n(\cdot)$, the smoothed empirical distribution function is defined by

$$\hat{F}_n^{sm}(x) := \frac{1}{n} \sum_{i=1}^{l} I_{(-\infty;x]}(x_i) + \frac{1}{n} \sum_{i=l+1}^{n} \frac{\int\limits_{-\infty}^{x} \xi_i(t)dt}{\int\limits_{-\infty}^{\infty} \xi_i(t)dt} \quad \forall x \in \mathbb{R}.$$

In Figure 6.3 a so-called mixed sample and the corresponding smoothed empirical distribution function are depicted.

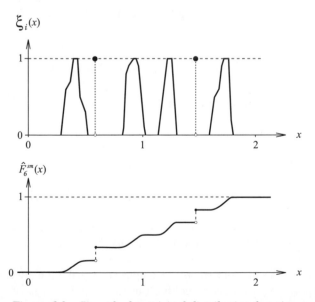

Figure 6.3   Smoothed empirical distribution function.

6.4 Problems

(a) Use the fuzzy sample from Figure 6.1 and draw diagrams of the corresponding smoothed empirical distribution function as well as the corresponding adapted cumulative sum function.

(b) Calculate the fuzzy 0.5 fractile for the fuzzy sample from Figure 6.1. Moreover make a diagram of the characterizing function of this fuzzy fractile.

7

Empirical correlation for fuzzy data

For classical two-dimensional data $(x_i, y_i) \in \mathbb{R}^2$, $i = 1(1)n$ the empirical correlation coefficient r is defined by

$$r := \frac{\sum\limits_{i=1}^{n} (x_i - \bar{x}_n)(y_i - \bar{y}_n)}{\sqrt{\left[\sum\limits_{i=1}^{n} (x_i - \bar{x}_n)^2 \right] \cdot \left[\sum\limits_{i=1}^{n} (y_i - \bar{y}_n)^2 \right]}}.$$

Remark 7.1: For r the following holds:

(a) $-1 \leq r \leq 1$.
(b) $|r| = 1$ if and only if all points (x_i, y_i) are on a line in the (x, y)-plane.

7.1 Fuzzy empirical correlation coefficient

For real data fuzziness can be present in different ways:

(a) $(x_i, y_i)^*$ can be fuzzy two-dimensional vectors.
(b) (x_i^*, y_i^*) are pairs of fuzzy numbers.
(c) One coordinate is precise and the other is fuzzy.

In order to apply the extension principle for the generalization of the empirical correlation coefficient, first the fuzzy data have to be combined into a fuzzy vector of

Statistical Methods for Fuzzy Data Reinhard Viertl
© 2011 John Wiley & Sons, Ltd

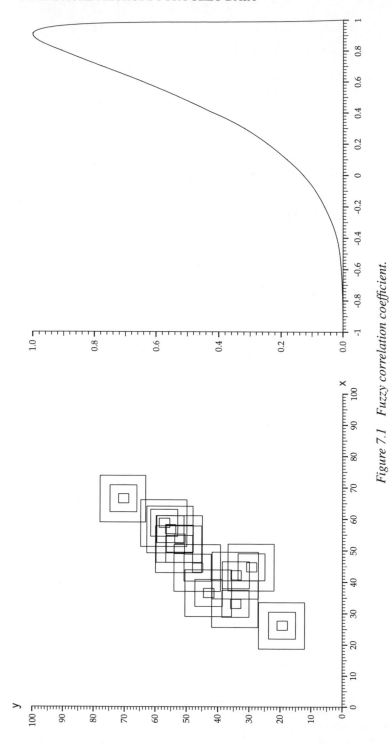

Figure 7.1   Fuzzy correlation coefficient.

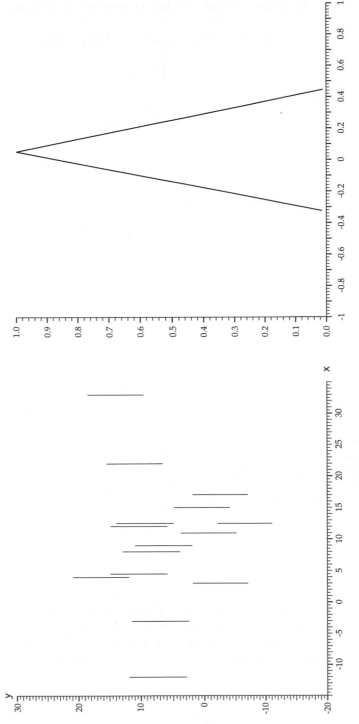

Figure 7.2   Partially fuzzy data and correlation coefficient.

the $2n$-dimensional Euclidean space $\mathbb{R}^{2n}$. For this combination the minimum t-norm is used.

In case (a) the fuzzy data consist of two-dimensional fuzzy vectors

$$\underline{x}_i^*, \quad i = 1(1)n$$

with corresponding vector-characterizing functions $\xi_i(., .), i = 1(1)n$.

The combined fuzzy sample is a $2n$-dimensional fuzzy vector whose vector-characterizing function $\xi(., \ldots, .)$ is given by its values

$$\xi(x_1, y_1, \ldots, x_n, y_n) = \min\{\xi_i(x_i, y_i) : i = 1(1)n\} \, \forall \, (x_1, y_1, \ldots, x_n, y_n) \in \mathbb{R}^n.$$

Applying the extension principle to the function

$$(x_1, y_1, \ldots, x_n, y_n) \to \mathrm{corr}(x_1, y_1, \ldots, x_n, y_n)$$

$$= \frac{\sum\limits_{i=1}^{n} (x_i - \bar{x}_n)(y_i - \bar{y}_n)}{\sqrt{\left[\sum\limits_{i=1}^{n} (x_i - \bar{x}_n)^2\right] \cdot \left[\sum\limits_{i=1}^{n} (y_i - \bar{y}_n)^2\right]}}$$

the characterizing function $\psi_{r^*}(\cdot)$ of the generalized (fuzzy) empirical correlation coefficient r^* is given by its values

$$\psi_{r^*}(r) := \begin{cases} \sup\{\xi(x_1, y_1, \ldots, x_n, y_n) : \text{for corr}(x_1, y_1, \ldots, x_n, y_n) = r\} \\ 0 \qquad \text{otherwise} \end{cases} \quad \forall r \in \mathbb{R}.$$

Remark 7.2: The support of r^* is a subset of the interval $[-1; 1]$.

In Figure 7.1 an example of fuzzy two-dimensional data and the corresponding fuzzy correlation coefficient is depicted.

Cases (b) and (c) are treated in a similar way. The only difference is that $\xi_i(., .)$ have to be formed by applying the minimum t-norm also. The rest is analogous to case (a).

An example for this case is given in Figure 7.2.

7.2 Problems

(a) What shape has the characterizing function of the empirical correlation coefficient in the case of data in the form of two-dimensional intervals?
(b) Explain in detail how to obtain the vector-characterizing function $\xi(., \ldots, .)$ of the combined fuzzy sample if the x-coordinates in $(x, y)_i^*$ are exact.

Part III

FOUNDATIONS OF STATISTICAL INFERENCE WITH FUZZY DATA

Statistical inference is based on stochastic models like probability distributions, parametric families of probability distributions, cumulative distribution functions, expectations, dependence structures, and others.

In this part the basic mathematical concepts for statistical inference in the case of fuzzy data as well as fuzzy probabilities are explained.

In standard statistical inference the combination of observations of a classical random variable to form an element of the sample space is trivial. Different from that for fuzzy data the combination is nontrivial because a vector of fuzzy numbers is not a fuzzy vector.

Therefore it is important to distinguish between observation space and sample space.

There are survey papers on different concepts concerned with statistical inference for fuzzy data by Gebhardt *et al.* (1997) and Taheri (2003).

8

Fuzzy probability distributions

Standard probability densities are usually motivated by histograms. Based on fuzzy histograms from Chapter 5 and by the fuzziness of a priori distributions in Bayesian inference a generalization of probability distributions is necessary. Moreover, a generalized law of large numbers for fuzzy valued sequences of random variables leads to so-called fuzzy probability densities.

8.1 Fuzzy probability densities

The theoretical counterparts of fuzzy histograms are fuzzy valued functions defined on observation spaces M, which are obeying the rules for fuzzy histograms (cf. Sections 3.6 and 5.2). These functions are called *fuzzy probability densities* (defined below).

Definition 8.1: A fuzzy valued function $f^* : M \to \mathcal{F}_I(\mathbb{R})$ defined on a measure space $(M, \mathcal{A}, \mu)$ is called a *fuzzy probability density* if it fulfills the following conditions:

1. The integral $y^* = \displaystyle\oint_M f^*(x)d\mu(x)$ defined in Section 3.6 exists.

2. $1 \in C_1(y^*)$ and $\mathrm{supp}(y^*) \subseteq (0, \infty)$.

3. There exists a classical probability density $f : M \to [0; \infty)$ with

$$\underline{f}_1(x) \le f(x) \le \overline{f}_1(x) \quad \text{for all} \quad x \in M.$$

Remark 8.1: Fuzzy probability densities can be graphically displayed by drawing δ-level functions $\underline{f}_\delta(\cdot)$ and $\overline{f}_\delta(\cdot)$ of $f^*(\cdot)$. An example of a one-dimensional fuzzy probability density, defined on the measure space $(\mathbb{R}, \mathcal{B}, \lambda)$, with λ denoting the Lebesgue measure, is depicted in Figure 8.1.

Statistical Methods for Fuzzy Data   Reinhard Viertl
© 2011 John Wiley & Sons, Ltd

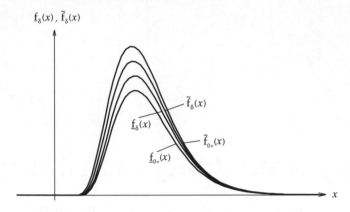

$$\underline{f}_\delta(x),\, \overline{f}_\delta(x)$$

Figure 8.1   One-dimensional fuzzy probability density.

8.2   Probabilities based on fuzzy probability densities

Based on the calculation of probabilities of events for classical continuous probability distributions by integrals, i.e.

$$P(A) = \int_A f(x)d\mu(x),$$

one could think to use the generalized integration from Definition 3.3 for fuzzy probability densities $f^*(\cdot)$.

However, this is not suitable because

$$\oint_M f^*(x)d\mu(x) \neq 1,$$

but as abstraction of the relative frequency $h_n^*(M)$ for fuzzy data $h_n^*(M) = 1$ has to hold true.

Therefore another generalized integration is necessary. This is possible by introducing the so-called *fuzzy probability integral*

$$(P)\int f^*(x)d\mu(x).$$

Definition 8.2: Let $f^*(\cdot)$ be a fuzzy probability density defined on a measure space $(M, \mathcal{A}, \mu)$, and consider the set S_δ of all classical probability densities $f(\cdot)$ on $(M, \mathcal{A}, \mu)$ obeying $\underline{f}_\delta(x) \leq f(x) \leq \overline{f}_\delta(x)$ for all $x \in M$.

Then the *fuzzy probability* $P^*(A)$ for $A \in \mathcal{A}$ is the fuzzy number whose δ-cuts $C_\delta[P^*(A)] = [\underline{P}_\delta(A); \overline{P}_\delta(A)]$ are defined for all $\delta \in (0; 1]$ by

$$\overline{P}_\delta(A) := \sup \left\{ \int_A f(x)d\mu(x) : f(.) \in S_\delta \right\}$$

and

$$\underline{P}_\delta(A) := \inf \left\{ \int_A f(x)d\mu(x) : f(.) \in S_\delta \right\}.$$

By the representation Lemma 2.1 the fuzzy number $P^*(A)$ is determined. It is denoted by

$$P^*(A) = (P) \int_A f^*(x)d\mu(x).$$

Remark 8.2: These generalized probability distributions P^* need not be additive.

Analogous to fuzzy relative frequencies (cf. Section 5.2) any such fuzzy probability distribution is fulfilling the following two conditions for disjoint events A and B and $\delta \in (0; 1]$:

$$\overline{P}_\delta(A \cup B) \leq \overline{P}_\delta(A) + \overline{P}_\delta(B)$$

and

$$\underline{P}_\delta(A \cup B) \geq \underline{P}_\delta(A) + \underline{P}_\delta(B).$$

The proof is given in Viertl and Hareter (2006).

An example is given in Section 8.4.

8.3   General fuzzy probability distributions

In abstraction of fuzzy histograms and generalizing fuzzy probability distributions generated by fuzzy probability densities, and also to include fuzzy discrete probabilities, the following definition is appropriate.

Definition 8.3: Let $(M, \mathcal{A})$ be a measurable space and $\mathcal{F}_I([0; \infty))$ be the set of all fuzzy intervals whose support is contained in $[0, \infty)$. Then a fuzzy probability distribution P^* on $(M, \mathcal{A})$ is a function $P^* : \mathcal{A} \to \mathcal{F}_I([0; \infty))$ obeying the following:

1. $P^*(\emptyset) = 0$.

2. $P^*(M) = 1$.

3. For all $\delta \in (0; 1]$ and pairwise disjoint events $A_1, \ldots, A_n$ in $\mathcal{A}$

$$\overline{P}_\delta \left(\bigcup_{i=1}^{n} A_i \right) \leq \sum_{i=n}^{n} \overline{P}_\delta(A_i)$$

and

$$\underline{P}_\delta \left(\bigcup_{i=1}^{n} A_i \right) \geq \sum_{i=n}^{n} \underline{P}_\delta(A_i).$$

Remark 8.3: This concept contains classical probability distributions as a special case.

8.4   Problems

(a) Consider a fuzzy uniform density $f^*(\cdot)$ on the interval $[0; 1]$ and calculate the fuzzy integral $\displaystyle\oint_{0}^{1} f^*(x)dx$ from Section 3.6.

(b) For the fuzzy density in (a) calculate the fuzzy probability $P^* \left([0; \tfrac{1}{2}] \right)$.

9

A law of large numbers

Random samples whose values are fuzzy exhibit also a kind of convergence in extension of the classical law of large numbers. A suitable mathematical model for random variables which assume fuzzy values are so-called fuzzy random variables. Based on this concept a generalized law of large numbers can be formulated.

9.1 Fuzzy random variables

Let $(\Omega, \mathcal{A}, P)$ be a classical probability space. The generalization of random variables X defined on $(\Omega, \mathcal{A}, P)$, i.e. X is a Borel-measurable function $X \colon \Omega \to \mathbb{R}$, to the situation of fuzzy valued quantities are fuzzy random variables, which are defined in the following way.

Definition 9.1: A *fuzzy random variable* X^* on a probability space $(\Omega, \mathcal{A}, P)$ is a function from Ω to the set of fuzzy numbers $\mathcal{F}(\mathbb{R})$ such that

$$\{\omega \in \Omega : X_\delta(\omega) \cap B \neq \phi\} \in \mathcal{A}$$

for every Borel-set B and every $\delta \in [0; 1]$, where $X_\delta(\omega)$ denotes the δ-cut of $X^*(\omega)$.

Remark 9.1: If the values $X^*(\omega)$ are all fuzzy intervals, then the set $X_\delta(\omega)$ is a compact interval for all $\delta \in [0; 1]$, i.e.

$$X_\delta(\omega) = \left[\underline{X}_\delta(\omega); \overline{X}_\delta(\omega)\right] \quad \forall \delta \in (0; 1].$$

X_δ is then an interval-valued function.

Lemma 9.1: Let $(\Omega, \mathcal{A}, P)$ be a probability space and $\mathcal{B}(\mathbb{R})$ denote the Borel-subsets of $\mathbb{R}$. Then a function $X^*; \Omega \to \mathcal{F}_I(\mathbb{R})$, where $\mathcal{F}_I(\mathbb{R})$ denotes the set of all fuzzy

Statistical Methods for Fuzzy Data   Reinhard Viertl
© 2011 John Wiley & Sons, Ltd

intervals, is a fuzzy random variable if and only if

$$\{\omega \in \Omega : X_\delta(\omega) \subseteq B\} \in \mathcal{A}$$

holds for every $B \in \mathcal{B}(\mathbb{R})$ and every $\delta \in [0; 1]$.

Proof: The lemma is a consequence of the following equality:

$$\begin{aligned}
\{\omega \in \Omega : X_\delta(\omega) \subseteq B\} &= \{\omega \in \Omega : X_\delta(\omega) \cap B^c = \phi\} \\
&= \Omega \setminus \{\omega \in \Omega : X_\delta(\omega) \cap B^c \neq \phi\} \\
&= \{\omega \in \Omega : X_\delta(\omega) \cap B^c \neq \phi\}^c
\end{aligned}$$

Theorem 9.1: Let $(\Omega, \mathcal{A}, P)$ be a complete probability space and $\mathcal{B}(\mathbb{R})$ the system of Borel-sets. Then a function $X^*; \Omega \to \mathcal{F}_I(\mathbb{R})$ is a fuzzy interval valued random variable if and only if $\underline{X}_\delta$ and $\overline{X}_\delta$ defined in Remark 9.1 are classical real valued random variables for every $\delta \in [0, 1]$, i.e.

$$\underline{X}_\delta^{-1}(B) = \{\omega \in \Omega : \underline{X}_\delta(\omega) \in B\} \in \mathcal{A}$$

and

$$\overline{X}_\delta^{-1}(B) = \{\omega \in \Omega : \overline{X}_\delta(\omega) \in B\} \in \mathcal{A}$$

holds for every $B \in \mathcal{B}(\mathbb{R})$.

Proof: If X^* is a fuzzy interval valued random variable then $\{\omega \in \Omega : x \in X_\delta(\omega)\} \in \mathcal{A}$ holds since $\{x\} \in \mathcal{B}(\mathbb{R})$ and $x \in X_\delta(\omega)$ is equivalent to $X_\delta(\omega) \cap \{x\} \neq \emptyset$ for every $x \in \mathbb{R}$ and $\delta \in (0; 1]$.

By assumption $\underline{X}_1(\omega) \leq \overline{X}_1(\omega)$ holds for P almost every $\omega \in \Omega$, and therefore a set $N \in \mathcal{A}$ exists with $P(N) = 0$ such that for all $\omega \in N^c$ the topological interior

$$\mathrm{int}\,(X_\delta(\omega)) \neq \emptyset.$$

Suppose G to be an open set in $\mathbb{R}$ and $\delta \in (0; 1]$ and $\{\omega \in \Omega : x \in X_\delta(\omega)\} \in \mathcal{A}$ for every $\delta \in (0; 1]$ and every $x \in \mathbb{R}$. Then the following holds true:

$$\begin{aligned}
&\{\omega \in \Omega : X_\delta(\omega) \cap G \neq \phi\} \\
&= \{\omega \in N : X_\delta(\omega) \cap G \neq \phi\} \cup \{\omega \in N^c : X_\delta(\omega) \cap G \neq \phi\}
\end{aligned}$$

By the completeness the first set is a measurable set. Therefore it is sufficient to prove that the second set is measurable:

$$\begin{aligned}
\{\omega \in N^c : X_\delta(\omega) \cap G \neq \phi\} &= \{\omega \in N^c : \mathrm{int}\,(X_\delta(\omega)) \cap G \neq \phi\} \\
&= \{\omega \in N^c : \mathrm{int}\,(X_\delta(\omega)) \cap G \cap \mathbb{Q} \neq \phi\}
\end{aligned}$$

where $\mathbb{Q}$ denotes the set of rational numbers. The last set can be written as

$$N^c \cap \left(\bigcup_{x \in \mathbb{Q} \cap G} \{\omega \in \Omega : x \in X_\delta(\omega)\} \right)$$

which is an element of $\mathcal{A}$. This completes the proof.

9.2   Fuzzy probability distributions induced by fuzzy random variables

Every fuzzy random variable naturally induces a fuzzy probability distribution. Let X^* be a fuzzy random variable defined on a probability space $(\Omega, \mathcal{A}, \mathrm{P})$. Defining the following functions $\underline{\pi}_\delta(\cdot)$ and $\overline{\pi}_\delta(\cdot)$ on the Borel σ-field $\mathcal{B}$:

$$\overline{\pi}_\delta(B) := \mathrm{P}(\{\omega \in \Omega : X_\delta(\omega) \cap B \neq \emptyset\})$$

and

$$\underline{\pi}_\delta(B) := \mathrm{P}(\{\omega \in \Omega : X_\delta(\omega) \subseteq B\}),$$

a fuzzy probability distribution P^* on the system of Borel-sets $\mathcal{B}$ is defined in the following way:

For $B \in \mathcal{B}$ the fuzzy interval $\mathrm{P}^*(B)$ is defined using the functions $\underline{\pi}_\delta(\cdot)$ and $\overline{\pi}_\delta(\cdot)$. The characterizing function $\xi(\cdot)$ of the fuzzy interval $\mathrm{P}^*(B)$ is defined by

$$\xi(x) := \begin{cases} \sup \left\{\delta \in (0;1] : x \in \left[\underline{\pi}_\delta(B); \overline{\pi}_\delta(B)\right]\right\} \\ 0 \ \ \text{if} \ \ x \notin \left[\underline{\pi}_\delta(B); \overline{\pi}_\delta(B)\right] \ \ \forall \delta \in (0;1] \end{cases} \quad \forall x \in \mathbb{R}.$$

The δ-cuts of the fuzzy interval $\mathrm{P}^*(B)$ are not necessarily identical to $[\underline{\pi}_\delta(B); \overline{\pi}_\delta(B)]$ and are denoted by

$$C_\delta \left[P^*(B)\right] = \left[\underline{P}_\delta(B); \overline{P}_\delta(B)\right] \quad \forall \delta \in (0;1].$$

The relationship between the functions above is given in the following lemma.

Lemma 9.2: Under the conditions of this section, let δ be an arbitrary fixed number in $[0;1]$. Suppose $(\delta_n)_{n \in \mathbb{N}}$ is a strictly increasing sequence in $(0;1)$ converging to δ.

Then the following holds for every Borel-set B:

$$
\begin{aligned}
\left[\underline{P}_\delta(B); \overline{P}_\delta(B)\right] &= \bigcap_{\beta < \delta} \left[\underline{\pi}_\beta(B); \overline{\pi}_\beta(B)\right] \\
&= \bigcap_{n=1}^{\infty} \left[\underline{\pi}_{\delta_n}(B); \overline{\pi}_{\delta_n}(B)\right] \\
&= \left[\lim_{n\to\infty} \underline{\pi}_{\delta_n}(B); \lim_{n\to\infty} \overline{\pi}_{\delta_n}(B)\right]
\end{aligned}
$$

The proof is given in Trutschnig (2006).

Based on Lemma 9.2 the following theorem can be proved.

Theorem 9.2: Suppose that $(\Omega, \mathcal{A}, P)$ is an arbitrary probability space and X^* is a fuzzy random variable defined on $(\Omega, \mathcal{A})$.

Let $[\underline{P}_\delta(B); \overline{P}_\delta(B)]$ be the δ-cuts of the fuzzy interval $P^*(B)$, i.e.

$$
\underline{P}_\delta(B) = \lim_{n\to\infty} \underline{\pi}_{\delta_n}(B) \text{ and } \overline{P}_\delta(B) = \lim_{n\to\infty} \overline{\pi}_{\delta_n}(B).
$$

Then P^* is a fuzzy probability distribution on the system of Borel-sets.

The proof is given in Trutschnig (2006).

9.3   Sequences of fuzzy random variables

Sequences of observations are basic for statistical inference. Moreover the law of large numbers makes sure that taking more observations is reasonable. Therefore it is necessary to consider sequences of fuzzy observations. Let $(X_n^*)_{n\in N}$ be a sequence of fuzzy random variables defined on a complete probability space $(\Omega, \mathcal{A}, P)$, such that for every $\omega \in \Omega$ the sequence $(X_n^*(\omega))_{n\in N}$ is cut-wise bounded.

In order to prove a generalized law of large numbers for fuzzy valued random variables, independence of fuzzy random variables has to be defined.

Definition 9.2: Suppose that $(\Omega, \mathcal{A}, P)$ is a complete probability space and X^*; $\Omega \to \mathcal{F}_N(\mathbb{R})$ and $Y^* : \Omega \to \mathcal{F}_N(\mathbb{R})$ are fuzzy random variables. Then X^* and Y^* are said to be independent if for arbitrary Borel-sets B_1 and B_2 the following equality holds for every $\delta \in (0; 1]$:

$$
P(X_\delta(\omega) \subseteq B_1, Y_\delta(\omega) \subseteq B_2) = P(X_\delta(\omega) \subseteq B_1) \cdot P(Y_\delta(\omega) \subseteq B_2).
$$

Proposition 9.1: Let $(\Omega, \mathcal{A}, P)$ be a complete probability space and X^* and Y^* independent fuzzy random variables. Then $\underline{X}_\delta$ and $\underline{Y}_\delta$ as well as $\overline{X}_\delta$ and $\overline{Y}_\delta$ are independent real-valued random variables.

The proof is given in Trutschnig (2006).

9.4   Law of large numbers for fuzzy random variables

In this section we consider sequences of pairwise independent, identically distributed fuzzy random variables $X_1^*, X_2^*, X_3^*, \ldots$. For such sequences fuzzy relative frequencies $h_n^*(B, \omega)$ are considered which are generated by the following family of sets:

$$\overline{h}_{n,\delta}(B, \omega) := \frac{1}{n} \# \left\{ i \in \{1, \ldots, n\} : X_{i,\delta}^*(\omega) \cap B \neq \phi \right\}$$

and

$$\underline{h}_{n,\delta}(B, \omega) := \frac{1}{n} \# \left\{ i \in \{1, \ldots, n\} : X_{i,\delta}^*(\omega) \subseteq B \right\}$$

For every $\omega \in \Omega$ and every Borel-set B the fuzzy relative frequencies $h_n^*(B, \omega)$ are defined as the convex hull of the family

$$\left(\left[\underline{h}_{n,\delta}(B, \omega); \overline{h}_{n,\delta}(B, \omega) \right] \right)_{\delta \in (0;1]}.$$

For the fuzzy relative frequencies the following lemma holds true.

Lemma 9.3: Suppose that $(\Omega, \mathcal{A}, P)$ is a complete probability space and $X_1^*, X_2^*, X_3^*, \ldots$ are pairwise independent, identically distributed fuzzy random variables, and B is a Borel-set. Let for every $\delta \in (0; 1]$ be $\underline{\pi}_\delta$ and $\overline{\pi}_\delta$ as in Section 9.2. Then for every $\delta \in (0; 1]$ there exists an event $N \in \mathcal{A}$ with $P(N) = 0$, such that for every $\omega \in N^c$ the following identities hold:

$$\lim_{n \to \infty} \overline{h}_{n,\delta}(B, \omega) = \overline{\pi}_\delta(B)$$

and

$$\lim_{n \to \infty} \underline{h}_{n,\delta}(B, \omega) = \underline{\pi}_\delta(B).$$

The proof is given in Trutschnig (2006).

The next step is to show that the limits can be used to define fuzzy intervals $[\underline{P}_\delta(B); \overline{P}_\delta(B)]$, $\delta \in (0; 1]$, which are defining fuzzy probabilities $P^*(B)$. This is the content of the following theorem.

Theorem 9.3: Suppose that $(\Omega, \mathcal{A}, P)$ is a complete probability space, that $X_1^*, X_2^*, X_3^*, \ldots$ is a sequence of pairwise independent, identically distributed fuzzy random variables and that B is an arbitrary Borel-set. Then there exists a set $\Lambda \subseteq [0; 1]$, fulfilling $\lambda(\Lambda) = 1$ (Lebesgue measure), and a set $N \in \mathcal{A}$, fulfilling $P(N) = 0$, such that for every $\delta \in \Lambda$ and every $\omega \in N^c$ the following holds:

$$\lim_{n \to \infty} \overline{h}_{n,\delta}(B, \omega) = \overline{P}_\delta(B)$$

and

$$\lim_{n \to \infty} h_{n,\delta}(B, \omega) = \underline{P}_\delta(B).$$

The proof is given in Trutschnig (2006).

Remark 9.2: Based on the above result different laws of large numbers, based on different metrics can be proved.

9.5 Problems

(a) Let $\psi(\cdot)$ be the membership function of a normalized fuzzy subset of $\mathbb{R}$ with bounded support. Find the convex hull of $\psi(\cdot)$. This is the characterizing function $\xi(\cdot)$ of the fuzzy number, such that $\psi(x) \leq \xi(x)\ \forall x \in \mathbb{R}$, and there is no characterizing function between $\psi(\cdot)$ and $\xi(\cdot)$.

(b) Prove the following: For every δ-cut of the characterizing function of a fuzzy number x^* the following is true:

$$C_\delta(x^*) = \bigcap_{\beta < \delta} C_\beta(x^*) \quad \forall \delta \in (0; 1].$$

10

Combined fuzzy samples

In standard statistics the combination of observations into an element of the sample space is trivial. But for fuzzy data this is not. Therefore some explanation is necessary.

10.1   Observation space and sample space

For a classical random variable X the set M_x of all possible values for X is called *observation space*, i.e.

$$\{M_x = \{X(\omega) : \omega \in \Omega\}.$$

Let $X_1, \ldots, X_n$ be a random sample of X, i.e. $X_1, \ldots, X_n$ are identically and independently distributed random variables having the same distribution as X. Then the set of all possible values which $(X_1, \ldots, X_n)$ can take is the Cartesian product of n copies of the observation space, i.e.

$$M_x \times M_x \times \cdots \times M_x = M_x^n,$$

called *sample space*.

If $x_1, \ldots, x_n$ is an observed sample from X then the observations x_i are elements of the observation space M_x, i.e. $x_i \in M_x$ for all $i = 1(1)n$.

Therefore the vector $(x_1, \ldots, x_n)$, which is the combined sample, is an element of the sample space, i.e. $(x_1, \ldots, x_n) \in M_x^n$. For precise observations this needs no further explanation. Moreover so-called statistics are functions of the sample, i.e. $s(x_1, \ldots, x_n)$, with $s : M_x^n \to N$, where N is a suitable set. The function $S = s(X_1, \ldots, X_n)$ is also called *statistic*.

For fuzzy data the situation is different, because of the fact that the observations are fuzzy numbers x_i^* with characterizing functions $\xi_i(\cdot)$. An observed fuzzy sample

Statistical Methods for Fuzzy Data   Reinhard Viertl
© 2011 John Wiley & Sons, Ltd

$x_1^*, \ldots, x_n^*$ consists of n fuzzy numbers. But these fuzzy numbers do not form a fuzzy element of the sample space M_x^n. So $(x_1^*, \ldots, x_n^*)$ is not a suitable object to generalize statistical procedures, using the extension principle (Definition 3.1). In order to apply the extension principle the fuzzy sample $x_1^*, \ldots, x_n^*$ has to be combined into a fuzzy vector $\underline{x}^*$ which is a fuzzy element of the sample space M_x^n. This combination is possible by application of a triangular norm.

10.2   Combination of fuzzy samples

Let $x_1^*, \ldots, x_n^*$ be n fuzzy elements of the observations space M_x, and $\xi_1(\cdot), \ldots, \xi_n(\cdot)$ the corresponding characterizing functions. In order to obtain a fuzzy element $\underline{x}^*$ of the sample space M_x^n, the vector-characterizing function $\xi(., \ldots, .)$ of $\underline{x}^*$ has to be constructed. The vector-characterizing function $\xi(., \ldots, .)$ is defined via its values $\xi(x_1, \ldots, x_n)$ for $x_i \in M_x$ in the following way:

Let T be a t-norm,

$$\xi(x_1, \ldots x_n) = T\left(\xi_1(x_1), \ldots, \xi_n(x_n)\right) \quad \forall (x_1, \ldots, x_n) \in M_X^n.$$

For statistical inference the most suitable t-norm is the minimum t-norm, because using this t-norm the fuzziness of data is propagated in generalized inference procedures. Therefore the vector-characterizing function $\xi(., \ldots, .)$ is given by its values

$$\xi(x_1, \ldots, x_n) = \min\left\{\xi_1(x_1), \ldots, \xi_n(x_n)\right\}$$

and the corresponding fuzzy vector $\underline{x}^*$ is called the *combined fuzzy sample*. This combined fuzzy sample is basic for the generalization of statistical procedures to fuzzy data. The imprecision of fuzzy data is propagated by functions of data, i.e. $s(x_1^*, \ldots, x_n^*) = s(\underline{x}^*)$.

10.3   Statistics of fuzzy data

In standard statistics basic for many inference procedures are functions of data $x_1, \ldots, x_n$, i.e. $s(x_1, \ldots, x_n)$, where $s : M_x^n \to N$ are measurable functions from the sample space to a suitable measurable space N. For a random sample $x_1, \ldots, x_n$ of a stochastic quantity (also called random variable), the stochastic quantity $s(X_1, \ldots, X_n)$ is called *statistic*.

In the case of fuzzy samples $x_1^*, \ldots, x_n^*$ also the values $s(x_1^*, \ldots, x_n^*)$ become fuzzy, and the fuzziness of the value $s(x_1^*, \ldots, x_n^*)$ is expressed by a membership function $\eta(\cdot)$ of a fuzzy element in N.

In order to obtain the membership function $\eta(\cdot)$ first the fuzzy sample has to be combined into a fuzzy element $\underline{x}^*$ of the sample space. Then the extension principle can be applied to obtain $\eta(\cdot)$.

For fuzzy data $x_1^*, \ldots, x_n^*$ with corresponding characterizing functions $\xi_1(\cdot), \ldots, \xi_n(\cdot)$ the values $\eta(y)$ of $s(x_1^*, \ldots, x_n^*)$ are given by

$$\eta(y) = \begin{cases} \sup \left\{ \xi(\underline{x}) : \underline{x} \in M_X^n, \ s(\underline{x}) = y \right\} & \text{if } s^{-1}(\{y\}) \neq \phi \\ 0 & \text{if } s^{-1}(\{y\}) = \phi \end{cases} \quad \forall y \in N,$$

where $\underline{x} = (x_1, \ldots, x_n)$ and

$$\xi(\underline{x}) = \xi(x_1 \ldots x_n) = \min \left\{ \xi_1(x_1, \ldots, x_n) \right\} \quad \forall (x_1, \ldots, x_n) \in M_x^n.$$

Remark 10.1: Frequently $M_x \subseteq \mathbb{R}$ and $N \subseteq \mathbb{R}$. More generally $M_x \subseteq \mathbb{R}^k$ and $N \subseteq \mathbb{R}^s$.

In general $\eta(\cdot)$ need not be a characterizing function. But for continuous functions $s : \mathbb{R}_n \to \mathbb{R}$ by Theorem 3.1 $\eta(\cdot)$ is a characterizing function of a fuzzy number.

Special cases of the function $s(X_1, \ldots, X_n)$ are statistical estimators, pivotal quantities, and test statistics. The generalization of these statistics will be considered in the following chapters.

10.4   Problems

(a) For which kind of samples is the observation space identical with the sample space?
(b) Prove that for n fuzzy intervals the combined fuzzy sample is a fuzzy vector.
(c) For a sample of two fuzzy numbers draw an axonometric diagram to explain the difference from (x_1^*, x_2^*) to $\underline{x}^*$.

Part IV

CLASSICAL STATISTICAL INFERENCE FOR FUZZY DATA

Classical statistical inference is based on the assumption that stochastic quantities X have an underlying true probability model P_0. In order to estimate P_0 usually a family $\mathcal{P}$ of possible probability distributions is considered, i.e. $\mathcal{P} = \{P : P \text{ possible distribution of } X\}$.

In the case of parametric families $\{P_\theta : \theta \in \Theta\}$ the generalization of estimators to the situation of fuzzy data is necessary, i.e. to construct estimations $\hat{\theta}^*$ for the true parameter θ_0 of the underlying probability distribution P_{θ_0} of X.

Next a generalization of confidence regions in the case of fuzzy data is given, and the resulting fuzzy confidence regions are typical examples of fuzzy sets.

Based on fuzzy samples also test statistics have to be adapted. The values of a test statistic in the case of fuzzy data become fuzzy numbers. Therefore test decisions are not as simple as in the standard situation of data in the form of numbers or vectors. The resulting fuzzy values of test statistics contain more information than in the classical situation because fuzzy values of a test statistic can make it clear that more data are needed in order to provide a well-based decision.

11

Generalized point estimators

In this chapter classical point estimators based on samples $x_1, \ldots, x_n$ of a stochastic quantity X are considered. Such estimators are functions $\vartheta : M^n \to N$, where M is the *observation space* of X (i.e. the set of all possible values of X) and M^n is the *sample space* for a sample of size n. $(N, \mathcal{Q})$ is a measurable space, where N is the set of all possible values for the quantity to be estimated.

Frequently the quantity to be estimated is a parameter θ of a parametric stochastic model $X \sim P_\theta, \theta \in \Theta$ or a quantile, or another characteristic value of the unknown distribution P_0 of X.

In the following suitable classical estimators are considered, fulfilling standard quality requirements such as unbiasedness, efficiency, maximum likelihood, consistency, and others.

11.1 Estimators based on fuzzy samples

In reality frequently the observed samples are n fuzzy numbers $x_1^*, \ldots, x_n^*$ with corresponding characterizing functions $\xi_1(\cdot), \ldots, \xi_n(\cdot)$. Therefore a classical estimator

$$\vartheta : M^n \to N$$

becomes a function $\vartheta(x_1^*, \ldots, x_n^*)$ of n fuzzy variables. Therefore by the extension principle (cf. Chapter 10) the value $\vartheta(x_1^*, \ldots, x_n^*)$ becomes fuzzy too. The characterizing function $\eta(\cdot)$ of the fuzzy value $\vartheta(x_1^*, \ldots, x_n^*)$ is given by using the combined fuzzy sample $\underline{x}^*$ with vector characterizing function $\xi : M^n \to [0; 1]$ as explained in Section 10.3. Therefore the characterizing function of $\hat{\theta}^* = \vartheta(x_1^*, \ldots, x_n^*)$ is given by its values $\eta(\theta)$ for all $\theta \in \Theta$.

Using the notation $\underline{x} = (x_1, \ldots, x_n) \in M^n$ the values $\eta(\theta)$ are given by

$$\eta(\theta) = \begin{cases} \sup\{\xi(\underline{x}) : \vartheta(\underline{x}) = \theta\} & \text{if } \exists \underline{x} \in M^n : \vartheta(\underline{x}) = \theta \\ 0 & \text{if } \nexists \underline{x} \in M^n : \vartheta(\underline{x}) = \theta \end{cases} \quad \forall \theta \in \Theta.$$

Remark 11.1: $\hat{\theta}^*$ is a fuzzy element in the parameter space Θ whose memberships function is $\eta(\cdot)$.

Example 11.1 Let X be an exponentially distributed stochastic quantity with density

$$f(x|\theta) = \frac{1}{\theta} e^{-\frac{x}{\theta}} I_{(0;\infty)}(x) \text{ with } \theta \in (0; \infty)$$

then the optimal estimator $\hat{\theta}$ for the true parameter θ_0 based on a classical sample $x_1, \ldots, x_n$ is given by

$$\hat{\theta} = \frac{1}{n} \sum_{i=1}^{n} x_i,$$

i.e. the optimal estimator is the sample mean $\bar{x}_n$.

Let $x_1^*, \ldots, x_n^*$ be a fuzzy sample whose characterizing functions are given in Figure 11.1.

Then the fuzzy estimator $\hat{\theta}^*$ for θ_0 is again a fuzzy number with trapezoidal characterizing function, as given in Figure 11.2.

Remark 11.2: In the case of the fuzzy data x_i^*, $i = 1(1)n$ are all fuzzy intervals with δ-cuts $C_\delta(x_i^*) = [\underline{x}_{i,\delta}, \overline{x}_{i,\delta}]$, $i = 1(1)n$, also the fuzzy value $\hat{\theta}^*$ is a fuzzy interval whose δ-cuts $C_\delta(\hat{\theta}^*)$ are given by

$$C_\delta(\hat{\theta}^*) = \left[\frac{1}{n} \sum_{i=1}^{n} \underline{x}_{i,\delta}; \frac{1}{n} \sum_{i=1}^{n} \overline{x}_{i,\delta} \right] \quad \forall \delta \in (0; 1] .$$

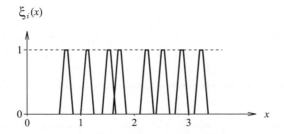

Figure 11.1   Fuzzy sample.

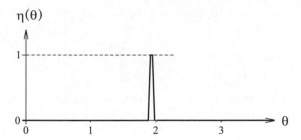

Figure 11.2   Characterizing function of the fuzzy estimator $\hat{\theta}^*$.

In the case where a transformed parameter $\tau(\theta_0)$ has to be estimated based on a fuzzy sample, the construction of the fuzzy estimator $\widehat{\tau(\theta_0)}^*$ is analogous.

Let $t : M^n \to \tau(\Theta)$ be a good classical estimator for $\tau(\theta_0)$, where

$$\tau(\Theta) := \{\tau(\theta) : \theta \in \Theta\}$$

is the space of transformed parameters. Then the membership function $\eta(\cdot)$ of the fuzzy estimator $\widehat{\tau(\theta)}^*$ based on the vector-characterizing function $\xi : M^n \to [0; 1]$ of the fuzzy combined sample $\underline{x}^*$ is given by its values $\eta(\tau)$ in the following way:

$$\eta(\tau) = \begin{cases} \sup \{\xi(\underline{x}) : t(\underline{x}) = \tau\} & \text{if } \exists \underline{x} : t(\underline{x}) = \tau \\ 0 & \text{if } \nexists \underline{x} : t(\underline{x}) = \tau \end{cases} \quad \forall \tau \in \tau(\Theta).$$

This is a consequence of the extension principle.

Remark 11.3: The construction in this chapter can be applied to all kinds of point estimators as long as reasonable estimators for the standard situation of precise samples $x_1, \ldots, x_n$ exist.

11.2   Sample moments

Without any parametric assumptions sample moments can be used to estimate moments of probability distributions.

The first sample moment $t_1(x_1, \ldots, x_n)$ is the sample mean

$$t_1(x_1, \cdots, x_n) = \bar{x}_n = \frac{1}{n} \sum_{i=1}^{n} x_i$$

which is used to estimate the first moment (i.e. the expectation) of a stochastic quantity.

The second centered sample moment $t_2(x_1, \ldots, x_n)$ is the sample variance

$$t_2(x_1, \ldots, x_n) = \frac{1}{n-1} \sum_{i=1}^{n} (x_i - \bar{x}_n)^2$$

which is used to estimate the variance of a stochastic quantity.

For $k > 2$ so-called higher sample moments m_k are defined by

$$m_k := \frac{1}{n} \sum_{i=1}^{n} x_i^k = t_k(x_1, \ldots, x_n).$$

In the case of fuzzy data $x_1^*, \ldots, x_n^*$ sample moments yield fuzzy values whose characterizing functions are given as in Section 11.1.

In the case where all fuzzy data are fuzzy intervals the δ-cuts of the sample moments are obtained by application of Theorem 3.1 since the sample moments are continuous functions of the data $x_1, \ldots, x_n$:

$$C_\delta(m_k^*) = \left[\min_{\underline{x} \in C_\delta(\underline{x}^*)} t_k(\underline{x}); \max_{\underline{x} \in C_\delta(\underline{x}^*)} t_k(\underline{x}) \right] \quad \forall \delta \in (0; 1].$$

For the sample mean $\bar{x}_n^* = m_1^*$ this reduces to the formula in Remark 11.2.

For the sample variance and its square root, called sample dispersion S_n^*, the calculation is more complicated.

For details, see Viertl and Hareter (2006).

11.3 Problems

(a) How can the characterizing function of the fuzzy estimate of the expectation of a one-dimensional stochastic quantity be calculated?

(b) How can the median of a one-dimensional distribution be estimated based on a fuzzy sample of size n?

12

Generalized confidence regions

Point estimators are based on qualitative goodness criteria, but often quantitative measures for the quality of an estimate are sought. Such estimates are *confidence regions*.

12.1 Confidence functions

In order to construct confidence regions (i.e. subsets $\Theta_{1-\alpha}$ of the parameter space Θ) which contain the true parameter θ_0 with high probability $1 - \alpha$ (where α is a very small positive number) so-called confidence functions are used. A confidence function

$$\kappa : M^n \to \mathcal{P}(\Theta)$$

is a function which assigns to every sample $x_1, \ldots, x_n$ a subset $\Theta_{1-\alpha}$ of Θ such that for a mathematical sample $X_1, \ldots, X_n$ of the observed stochastic quantity X the following is true:

$$P_\theta \{\theta \in \Theta_{1-\alpha} = \kappa(x_1, \ldots, x_n)\} = 1 - \alpha \quad \forall \theta \in \Theta.$$

For α a value of 0.05 or 0.01 is usually taken. Therefore the so-called coverage probability is $1 - \alpha$.

Statistical Methods for Fuzzy Data   Reinhard Viertl
© 2011 John Wiley & Sons, Ltd

12.2   Fuzzy confidence regions

In the case of fuzzy samples $x_1^*, \ldots, x_n^*$ of a parametric stochastic model $X \sim P_\theta, \theta \in \Theta$, for given confidence function $\kappa : M^n \to \mathcal{P}(\Theta)$ a *generalized confidence region* for the true parameter is the fuzzy subset $\Theta_{1-\alpha}^*$ of Θ, whose membership function $\varphi(\cdot)$ is defined in the following way:

$$\varphi(\theta) := \begin{cases} \sup\left\{\xi(\underline{x}) : \theta \in \kappa(\underline{x})\right\} & \text{if } \exists \underline{x} \in M^n : \theta \in \kappa(\underline{x}) \\ 0 & \text{if } \nexists \underline{x} \in M^n : \theta \in \kappa(\underline{x}) \end{cases} \quad \forall \theta \in \Theta$$

where $\underline{x} = (x_1, \ldots, x_n)$ and $\xi(\cdot)$ in the vector-characterizing function of the combined fuzzy sample $\underline{x}^*$ and M is the observation space of X.

Remark 12.1: The construction of the fuzzy confidence region is not an application of the extension principle. It should be noted that for classical data $x_1, \ldots, x_n$, taking the one-point indicator functions $I_{\{x_i\}}(\cdot)$, the resulting membership function of the generalized confidence region is the indicator function of the classical confidence region $\kappa(x_1, \ldots, x_n)$, i.e. $\varphi(\cdot) = I_{\kappa(x_1, \ldots, x_n)}(\cdot)$.

Remark 12.2: For the membership function $\varphi(\cdot)$ of the fuzzy confidence regions from this section under the conditions above and using the notation from before, the following holds:

$$I_{\underset{\underline{x}:\xi(\underline{x})=1}{\bigcup}\kappa(\underline{x})}(\theta) \leq \varphi(\theta) \quad \forall \theta \in \Theta.$$

To see this consider $\sup\left\{\xi(\underline{x}) : \theta \in \kappa(\underline{x})\right\}$.

Example 12.1 Let $X \sim Ex_\theta, \theta \in (0; \infty)$ be the family of exponential distributions, and $x_1^*, \ldots, x_n^*$ be a fuzzy sample of X.

The classical confidence function $\kappa(\underline{x}) = \kappa(x_1, \ldots, x_n)$ is

$$\kappa(x_1, \ldots, x_n) = \left[\frac{2\sum\limits_{i=1}^{n} X_i}{\chi^2_{2n;1-\frac{\alpha}{2}}} ; \frac{2\sum\limits_{i=1}^{n} X_i}{\chi^2_{2n;\frac{\alpha}{2}}} \right],$$

where $\chi^2_{2n;p}$ is the p-fractile of the chi-square distribution with $2n$ degrees of freedom. From this and using the definition of $\varphi(\cdot)$, the membership function can be obtained based on the vector-characterizing function $\xi(\cdot)$ of the fuzzy combined sample $\underline{x}^*$. For the fuzzy sample whose characterizing functions are given in Figure 12.1, the membership function of the fuzzy confidence interval for θ is depicted in Figure 12.2.

Example 12.2 Let $X \sim N(\mu, \sigma^2), \theta = (\mu, \sigma^2) \in \mathbb{R} \times (0; \infty)$ be the family of normal distributions. For classical sample $x_1, \ldots, x_n$ of X a two-dimensional confidence

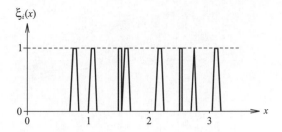

Figure 12.1   Fuzzy sample.

region for θ with coverage probability $1 - \alpha$ is given by the following two equations:

$$(\mu - \overline{x}_n)^2 = \frac{\sigma^2}{n} \cdot u_{(1-\sqrt{1-\alpha})/2}$$

$$\frac{(n-1)s_n^2}{\chi^2_{n-1;(1+\sqrt{1-\alpha})/2}} \quad \text{and} \quad \frac{(n-1)s_n^2}{\chi^2_{n-1;(1-\sqrt{1-\alpha})/2}}$$

Here u_p is the p-fractile of the standard normal distribution and $\chi^2_{n-1;p}$ is the p-fractile of the chi-square distribution with n-1 degrees of freedom.

For fuzzy samples using the definition of the membership function $\varphi(\cdot)$ of the fuzzy confidence region for θ above, the generalized confidence region $\Theta^*_{1-\alpha}$ can be obtained. For details, see Dutter and Viertl (1998). An example is depicted in Figure 12.3.

Proposition 12.1: Let $\underline{x}^*$ be a combined fuzzy sample of size n with vector-characterizing function $\xi : \mathbb{R}^n \to [0; 1]$ whose δ-cuts are simply connected, and $\kappa(x_1, \ldots, x_n)$ be a confidence function for the one-dimensional parameter θ of a stochastic model $X \sim P_\theta, \theta \in \Theta \subseteq \mathbb{R}$ with

$$\kappa(x_1, \ldots, x_n) = \left[\underline{\kappa}(x_1, \ldots, x_n); \overline{\kappa}(x_1, \ldots, x_n)\right],$$

where $\underline{\kappa}(\ldots)$ and $\overline{\kappa}(\ldots)$ are continuous functions. Then the membership function $\varphi(\cdot)$ of the fuzzy confidence region defined in this section is the characterizing function

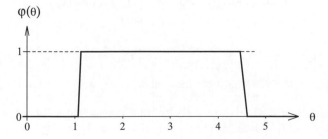

Figure 12.2   Fuzzy confidence interval.

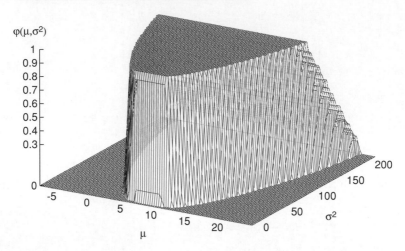

Figure 12.3   Fuzzy confidence region for the parameter vector of the normal distri-
bution.

of a fuzzy interval θ^* whose δ-cuts obey the following:

$$C_\delta(\theta^*) = \left[\min_{x \in C_\delta(\underline{x}^*)} \underline{\kappa}(x); \max_{x \in C_\delta(\underline{x}^*)} \overline{\kappa}(x) \right] \quad \forall \delta \in (0; 1].$$

Proof: By the continuity of $\underline{\kappa}(\ldots)$ and $\overline{\kappa}(\cdots)$ the set

$$\kappa^{-1}(\{\theta\}) := \left\{ \underline{x} : \theta \in \kappa(\underline{x}) \right\} = \left\{ \underline{x} : \underline{\kappa}(\underline{x}) \le \theta \le \overline{\kappa}(\underline{x}) \right\}$$

is closed.

The next step is to prove

$$C_\delta(\theta^*) = \bigcup_{\underline{x} \in C_\delta(\underline{x}^*)} \kappa(\underline{x}) = \bigcup_{\underline{x} \in C_\delta(\underline{x}^*)} \left[\underline{\kappa}(\underline{x}); \overline{\kappa}(\underline{x}) \right].$$

First,   let $\theta \in \bigcup_{\underline{x} \in C_\delta(\underline{x}^*)} \kappa(\underline{x}) \Rightarrow \exists \underline{x}_0 \in C_\delta(\underline{x}^*) : \theta \in \kappa(\underline{x}_0) \Rightarrow \xi(\underline{x}_0) \ge \delta \Rightarrow$
$\sup_{\theta \in \kappa(\underline{x})} \xi(\underline{x}) \ge \delta \Rightarrow \theta \in C_\delta(\theta^*).$ On   the   other   side   for $\theta \in C_\delta(\theta^*)$ we   have
$\varphi(\theta) = \sup \left\{ \xi(\underline{x}) : \theta \in \kappa(\underline{x}) \right\} \ge \delta$ and   by   the   compactness   of $\kappa(\underline{x})$ we   have
$\sup \left\{ \xi(\underline{x}) : \theta \in \kappa(\underline{x}) \right\} = \max \left\{ \xi(\underline{x}) : \theta \in \kappa(\underline{x}) \right\} \ge \delta \Rightarrow \exists \underline{x}_0 : \xi(\underline{x}_0) \ge \delta$ and
$\theta \in \kappa(\underline{x}_0)$ and therefore $\theta \in \bigcup_{\underline{x} \in C_\delta(\underline{x}^*)} \kappa(\underline{x}).$

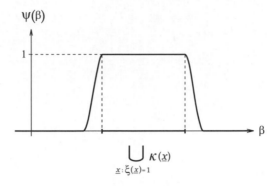

Figure 12.4   Fuzzy confidence interval.

In order to prove $\bigcup_{\underline{x} \in C_\delta(\underline{x}^*)} \left[\underline{\kappa}(\underline{x}); \overline{\kappa}(\underline{x})\right] = \left[\min_{\underline{x} \in C_\delta(\underline{x}^*)} \underline{\kappa}(\underline{x}); \max_{\underline{x} \in C_\delta(\underline{x}^*)} \overline{\kappa}(\underline{x})\right]$ by the conti-

nuity of $\underline{\kappa}(\ldots)$ and $\overline{\kappa}(\cdots)$ and the connectedness of $C_\delta(\underline{x}^*)$ $\forall \delta \in (0; 1]$ it follows that $\bigcup_{\underline{x} \in C_\delta(\underline{x}^*)} \left[\underline{\kappa}(\underline{x}); \overline{\kappa}(\underline{x})\right]$ is connected and compact and therefore a closed interval.

The concept of fuzzy confidence regions applies also for *transformed parameters* or *reduced parameters*. Let $\tau(\theta)$ be a function of the parameter θ of a stochastic model $X \sim P_\theta, \theta \in \Theta$ and $\tau(\Theta) = \{\tau(\theta) : \theta \in \Theta\} = B$ the set of all possible values of the transformed parameter. Moreover let $\kappa(x_1, \ldots, x_n)$ be a classical confidence function for $\beta = \tau(\theta)$ with given confidence level.

In the case of fuzzy data $x_1^*, \ldots, x_n^*$ of X with combined fuzzy sample $\underline{x}^*$ whose vector-characterizing function is $\xi(\cdot)$ the membership function $\psi(\cdot)$ of the generalized (fuzzy) confidence region for the transformed parameter β is defined by

$$\psi(\beta) = \begin{cases} \sup\{\xi(\underline{x}) : \beta \in \kappa(\underline{x})\} & \text{if } \exists \underline{x} : \tau(\theta) \in \kappa(\underline{x}) \\ 0 & \text{if } \nexists \underline{x} : \tau(\theta) \in \kappa(\underline{x}) \end{cases} \quad \forall \beta \in B.$$

Remark 12.3: Usually the transformed parameter is one-dimensional, i.e. $\tau(\theta) \in \mathbb{R}$.

A typical example is $X \sim N(\mu, \sigma^2)$ with $\theta = (\mu, \sigma^2)$ and $\tau(\theta) = \mu$.

For fuzzy confidence regions for transformed parameters similar to Remark 12.2 also here we have

$$I_{\bigcup_{\underline{x}:\xi(\underline{x})=1} \kappa(\underline{x})}(\beta) \leq \psi(\beta) \quad \forall \beta \in \tau(\Theta).$$

This is depicted for one-dimensional β in Figure 12.4.

12.3   Problems

(a) Calculate the fuzzy confidence interval for the parameter θ in Example 12.1.

(b) Prove the assertion in Remark 12.2.

13

Statistical tests for fuzzy data

Statistical test decisions are frequently based on measurable functions $t(x_1, \ldots, x_n)$ of observed samples $x_1, \ldots, x_n$.

Denoting by M_T the set of all possible values of the test statistic $T = t(X_1, \ldots, X_n)$, for hypothesis $\mathcal{H}$ concerning the distribution of the observed stochastic quantity X, the set M_T is usually decomposed into a *region of acceptance* of $\mathcal{H}$, denoted by A, and a so-called *critical region* C, i.e. $M_T = A \cup C$ with $A \cap C = \emptyset$. The decision rule is the following.

Let $x_1, \ldots, x_n$ be the observed sample, then:

if $t(x_1, \ldots, x_n) \in A \Rightarrow \mathcal{H}$ is accepted if $t(x_1, \ldots, x_n) \in C = M_T \backslash A \Rightarrow \mathcal{H}$ is rejected.

The subsets A and C are constructed using the probability α of an error of the first type, i.e. the probability of rejecting a true hypothesis $\mathcal{H}$:

$$\alpha = \Pr\{\mathcal{H} \text{ is rejected} | \mathcal{H} \text{ is true}\}$$

For one-dimensional test statistics T critical values t_{crit} for T are calculated and the test decision is based on the comparison of t_{crit} with the observed value $t = t(x_1, \ldots, x_n)$.

In this situation a decision is always possible.

13.1 Test statistics and fuzzy data

For fuzzy samples $x_1^*, \ldots, x_n^*$ of a stochastic quantity X the value of a test statistic $t(x_1^*, \ldots, x_n^*)$ becomes fuzzy too. The characterizing function $\psi(\cdot)$ of the fuzzy value $t^* = t(x_1^*, \ldots, x_n^*)$ is obtained by the extension principle (cf. Section 10.3).

Using the notation from the beginning of this chapter there are three possible situations:

(1) The support supp(t^*) is a subset of A.

(2) The support supp(t^*) is a subset of C.

(3) supp (t^*) has non-empty intersections with both A and C.

Situations (1) and (2) yield a decision as in the case of precise observations (which are usually unrealistic for continuous quantities X).

In situation (3) a decision is not possible because one cannot decide if the value t^* of the test statistic is in the acceptance region or not.

This situation indicates that the information contained in the given sample is not sufficient in order to make a well-based decision.

The three possible situations are displayed in Figure 13.1. Cases (1) and (2) are supporting a test decision. Case (3) needs more data in order to come to a decision.

For two-sided test problems the situations are analog.

In applications often p-values are calculated for making test decisions.

For fuzzy values t^* of a test statistic $t(x_1, \ldots, x_n)$ p-values can be obtained in the following way: Based on p-values for classical values $t = t(x_1, \ldots, x_n)$ of a test statistic, i.e. the p-value of t corresponding to a hypothesis $\mathcal{H}$ is that probability of an error of the first type (i.e. the probability of rejecting $\mathcal{H}$ if $\mathcal{H}$ is true) for which based on the observed t the change from acceptance of $\mathcal{H}$ to the rejection of $\mathcal{H}$ takes place.

For fuzzy value t^* of the test statistic with characterizing function $\xi_{t^*}(\cdot)$ the corresponding p-value is defined to be that probability of an error of the first type, for which the support of $\xi_{t^*}(\cdot)$ is just contained in the rejection region of the corresponding classical test. An example is depicted in Figure 13.2.

The p-value corresponding to t^* is the p-value which is obtained for the classical value t_0 of the test statistic.

Remark 13.1: Fuzzy values of a test statistic are realistic information concerning test problems. They give valuable hints when additional data are necessary to make a well-based decision.

13.2   Fuzzy p-values

Let $\eta(\cdot)$ be the characterizing function of the fuzzy value $t^* = t(x_1^*, \ldots, x_n^*)$ of a test statistic $T = t(x_1, \ldots, x_n)$ based on the fuzzy sample $x_1^*, \ldots, x_n^*$.

In applications the support supp $[\eta(\cdot)] = \{x \in \mathbb{R} : \eta(x) > 0\}$ is usually a bounded subset of $\mathbb{R}$.

Considering the δ-cuts of t^* for $\delta \in (0; 1]$ and denoting them by

$$C_\delta(t^*) = [t_1(\delta); t_2(\delta)] \quad \forall \delta \in (0; 1]$$

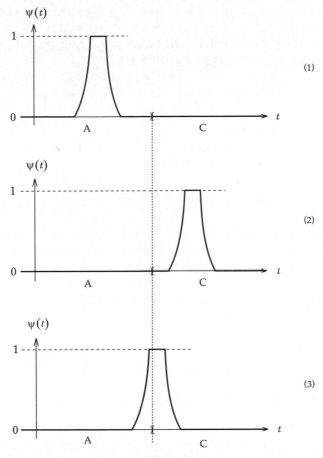

$\psi(\cdot)$ is the characterizing function of t^*.

Figure 13.1   Testing situations with fuzzy samples.

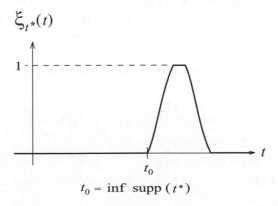

$t_0 = \inf \, \text{supp} \, (t^*)$

Figure 13.2   p-value of a fuzzy value of a test statistic.

the corresponding *fuzzy p-value* p^* is defined via its δ-cuts $C_\delta(p^*)$ for $\delta \in (0; 1]$ in the following way:

For a *one-sided test* with decision rule for $T \leq t_{cr}$ reject hypothesis, the δ-cuts of the corresponding fuzzy p-value p^* are defined by

$$C_\delta(p^*) = [\Pr\{T \leq t_1(\delta)\}; \Pr\{T \leq t_2(\delta)\}] \quad \forall \delta \in (0; 1]$$

with $t_1(\delta)$ and $t_2(\delta)$ from above.

For one-sided tests with decision rule for $T \geq t_{cr}$ reject the hypothesis, the δ-cults of the corresponding fuzzy p-value p^* are defined by

$$C_\delta(p^*) = [\Pr\{T \geq t_2(\delta)\}; \Pr\{T \geq t_1(\delta)\}].$$

In the case of *two-sided tests* first it has to be decided on which side of the median m of the distribution of T the main part of the amount of fuzziness of t^* is located. Therefore one has to compute the area under the characterizing function $\eta(\cdot)$ of t^* which is on the left side of m and the area on the right side of m.

Denoting these areas by A_1 and A_2, respectively, the δ-cuts of the corresponding fuzzy p-value p^* are defined by

$$C_\delta(p^*) = \begin{cases} [2\Pr\{T \leq t_1(\delta)\}; \; \min\{1, \Pr\{T \leq t_2(\delta)\}] & \text{if } A_1 > A_2 \\ [2\Pr\{T \geq t_2(\delta)\}; \; \min\{1, \Pr\{T \geq t_1(\delta)\}] & \text{if } A_1 \leq A_2 \end{cases} \quad \forall \delta \in (0; 1].$$

Proposition 13.1: The intervals $C_\delta(p^*)$ defined above are δ-cuts of a characterizing function.

Proof: Since t^* is a fuzzy number with characterizing function $\eta(\cdot)$, the intervals $[t_1(\delta); t_2(\delta)]$ are δ-cuts for all $\delta \in (0; 1]$. If t^* is a fuzzy interval then the δ-cuts $C_\delta(p^*)$ are closed finite intervals $[p_1(\delta); p_2(\delta)]$.

Therefore in this case p^* is a fuzzy interval.

Note that $p_1(\delta) \geq 0$ and $p_2(\delta) \leq 1$ for all $\delta \in (0; 1]$. Therefore the δ-cuts of p^* can be interpreted in terms of probabilities and compared with the significance level α of the test.

Now we have the following decision rule for fuzzy values t^* of the test statistic T:

If, for all $\delta \in (0; 1]$ and $p_1(\delta) \leq p_2(\delta)$

$p_1(\alpha) < \alpha$: reject $\mathcal{H}$

$p_2(\alpha) > \alpha$: accept $\mathcal{H}$

for $\alpha \in [p_1(\delta); p_2(\delta)]$ take more observations.

Remark 13.2: In the third case the uncertainty of making a decision is expressed.

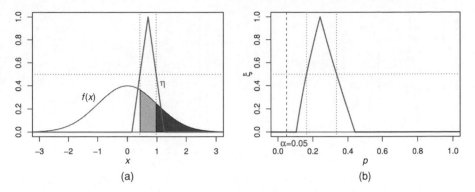

Figure 13.3   Construction of the characterizing function of the fuzzy p-value.

Example 13.1 Let T have a standard normal distribution, and $\eta(\cdot)$ a triangular characterizing function of the fuzzy value t^* of the test statistic with center 0.7. Figure 13.3(a) shows the density $f(\cdot)$ of the standard normal distribution and the characterizing function $\eta(\cdot)$.

In order to test the hypothesis

$$\mathcal{H}_0: \theta \leq \theta_0$$

against the alternative

$$\mathcal{H}_1: \theta > \theta_0 \text{ taking } \theta_0 = 0$$

we have a one-sided test. The fuzzy p-value p^* is determined as above. The characterizing function of p^* is presented in Figure 13.3(b).

Figure 13.3 shows in detail how the δ-cut of p^* is computed for $\delta = 0.5$.

In Figure 13.3(a) the δ-cut $C_\delta(t^*) = [t_1(\delta); t_2(\delta)]$ for $\delta = 0.5$ is indicated by two vertical lines. The exact p-value for t_2 (0.5) is shown by the dark area, and for t_1 (0.5) by the dark and gray area und the function $f(\cdot)$. These p-values form the δ-cut $C_\delta(p^*)$ for $p = 0.5$, which is presented in Figure 13.3(b) at the intersection of the horizontal line with the vertical lines through the exact p-values. Using this procedure for all $\delta \in (0; 1]$, the characterizing function $\xi(\cdot)$ of p^* is obtained. Finally, the resulting fuzzy p-value is compared with the significance level α which has to be fixed in advance. If supp$[p^*]$ does not contain α a decision is possible. Otherwise more data have to be taken in order to make a well-based decision.

Example 13.2 Continuing Example 13.1 and assuming the characterizing function $\eta(\cdot)$ of t^* to be of triangular shape, we consider four cases as depicted in Figure 13.4(a). For the four characterizing functions $\eta_1, \ldots, \eta_4(\cdot)$ the characterizing functions $\xi_1(\cdot), \ldots, \xi_4(\cdot)$ of the corresponding fuzzy p-values are given in Figure 13.4(b).

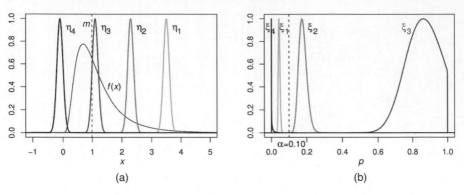

$f(\cdot)$ is the standard normal density. $\mathcal{H}_0$ is rejected for $\eta_1(\cdot)$, not rejected for $\eta_3(\cdot)$ until $\eta_4(\cdot)$. No decision is possible for $\eta_2(\cdot)$, all at significance level $\alpha = 0.5$.

Figure 13.4   Fuzzy p-values for different fuzzy values t^* of a test statistic.

Remark 13.3: For fuzzy p-values in the case of fuzzy hypotheses, see Parchami *et al.* (2010).

Further details are given in Filzmoser and Viertl (2004).

13.3   Problems

(a) Take a fuzzy sample of a normal distribution $N(\mu, \sigma^2)$ whose characterizing functions are trapezoidal. How is the p-value for a hypothesis $\mathcal{H}: \mu = \mu_0$ with given μ_0 calculated?

(b) Work out the details of the construction of the characterizing function of p^* in Figure 13.3.

(c) Continuing from (a), how is the fuzzy p-value p^* calculated?

Part V

BAYESIAN INFERENCE AND FUZZY INFORMATION

In standard Bayesian statistics all unknown quantities are described by stochastic quantities and their probability distributions. Consequently, so are the parameters θ of stochastic models $X \sim f(\cdot|\theta)$; $\theta \in \Theta$ in the continuous case, and $X \sim p(\cdot|\theta)$; $\theta \in \Theta$ in the discrete case. The a priori knowledge concerning θ is expressed by a probability distribution $\pi(\cdot)$ on the parameter space Θ, called *a priori distribution*.

If the parameter θ is continuous, the a priori distribution has a density function, called *a priori density*.

For discrete parameter space $\Theta = \{\theta_1, \ldots, \theta_m\}$ the a priori distribution is a discrete probability distribution with point probabilities $\pi(\theta_j)$, $j = 1(1)m$.

If the observed stochastic quantity $X \sim p(\cdot|\theta)$; $\theta \in \{\theta_1, \cdots, \theta_m\}$ is also discrete, i.e. the observation space M_X of X is countable with

$$M_X = \{x_1, x_2, \ldots\}$$

and observations $\overset{\circ}{x}_1, \ldots, \overset{\circ}{x}_n$ of X are given [sample data $D = (\overset{\circ}{x}_1, \ldots, \overset{\circ}{x}_n)$], then the conditional distribution $\Pr(\widetilde{\theta}|D)$ of the parameter is obtained via Bayes' formula in the following way: The so-called *a posteriori probabilities* $\pi(\theta_j|D)$ of the parameter values θ_j, given the data $D = (\overset{\circ}{x}_1, \ldots, \overset{\circ}{x}_n)$, are obtained by

$$\pi\left(\theta_j|\overset{\circ}{x}_1, \ldots, \overset{\circ}{x}_n\right) = \Pr\left\{\widetilde{\theta} = \theta_j|X_1 = \overset{\circ}{x}_1, \ldots, X_n = \overset{\circ}{x}_n\right\}$$

$$= \frac{\Pr\left\{X_1 = \overset{\circ}{x}_1, \ldots, X_n = \overset{\circ}{x}_n, \widetilde{\theta} = \theta_j\right\}}{\Pr\left\{X_1 = \overset{\circ}{x}_1, \ldots, X_n = \overset{\circ}{x}_n\right\}}$$

$$= \frac{p\left(\overset{\circ}{x}_1, \ldots, \overset{\circ}{x}_n | \theta_j\right) \cdot \pi(\theta_j)}{\sum\limits_{j=1}^{m} p\left(\overset{\circ}{x}_1, \ldots, \overset{\circ}{x}_n | \theta_j\right) \cdot \pi(\theta_j)} \qquad \forall \theta_j \in \Theta.$$

In the case of independent observations $\overset{\circ}{x}_1, \ldots, \overset{\circ}{x}_n$ the joint probability $p\left(\overset{\circ}{x}_1, \ldots, \overset{\circ}{x} | \theta_j\right)$ is given by

$$p(\overset{\circ}{x}_1, \ldots, \overset{\circ}{x}_n | \theta_j) = \prod_{i=1}^{n} p(\overset{\circ}{x}_i | \theta_j).$$

This probability distribution on the parameter space $\Theta = \{\theta_1, \ldots, \theta_m\}$ is called *a posteriori distribution* of $\widetilde{\theta}$, where $\widetilde{\theta}$ denotes the stochastic quantity describing the uncertainty concerning the parameter.

For continuous stochastic quantities X with parametric probability density $f(.|\theta)$ and continuous parameter space $\Theta \subseteq R^k$, i.e. stochastic model

$$X \sim f\left(\cdot|\theta\right); \; \theta \in \Theta,$$

the a priori distribution of $\widetilde{\theta}$ is given by a probability density $\pi(\cdot)$ on the parameter space Θ. This probability density is called *a priori density* which obeys

$$\int_{\Theta} \pi(\theta)d\theta = 1.$$

For observed sample $x_1, \ldots, x_n$ of X the so-called *a posteriori density* $\pi\left(\cdot|x_1, \cdots, x_n\right)$ is the conditional density of the parameter, given the data

$x_1, \ldots, x_n$, i.e.

$$\pi\left(\theta|x_1, \ldots, x_n\right) = \frac{g\left(x_1, \ldots, x_n, \theta\right)}{h\left(x_1, \ldots, x_n\right)} \qquad \forall \theta \in \Theta$$

where $g(., \ldots, .)$ is the joint probability density of the stochastic vector $\left(X_1, \ldots, X_n, \widetilde{\theta}\right)$ and $h(x_1, \ldots, x_n)$ is the marginal density of $(X_1 = x_1, \ldots, X_n = x_n)$.

In the case of independent observations $x_1, \ldots, x_n$ the joint density $g(x_1, \ldots, x_n, \theta)$ is given by the following product:

$$g\left(x_1, \ldots, x_n, \theta\right) = \left[\prod_{i=1}^{n} f\left(x_i|\theta\right)\right] \cdot \pi(\theta).$$

The function $\prod_{i=1}^{n} f\left(x_i|\theta\right) = l\left(\theta; x_1, \ldots, x_n\right)$ is the *likelihood function*.

The marginal density $h(x_1, \ldots, x_n)$ is given by

$$h(x_1, \ldots, x_n) = \int_\Theta l(\theta; x_1, \ldots, x_n) \cdot \pi(\theta) d\theta.$$

Therefore the a posteriori density is given by its values

$$\pi(\theta|x_1, \ldots, x_n) = \frac{\pi(\theta) \cdot l(\theta; x_1, \ldots, x_n)}{\int_\Theta \pi(\theta) \cdot l(\theta; x_1, \ldots, x_n) d\theta} \quad \forall \theta \in \Theta.$$

This is called *Bayes' theorem.*

Remark V.1: After the sample $x_1, \ldots, x_n$ is observed these values are constants. Therefore $h(x_1, \ldots, x_n)$ is a normalizing constant, and Bayes' theorem can be written in its short form

$$\pi(\theta|x_1, \ldots, x_n) \propto \pi(\theta) \cdot l(\theta; x_1, \ldots, x_n)$$

where $\propto$ stands for 'proportional to $\pi(\theta) \cdot l(\theta x_1, \ldots, x_n)$ up to a normalizing constant'.

For more general data D – for example for censored data – the likelihood function has to be adapted. The general form of Bayes' theorem is given by

$$\pi(\theta|D) \propto \pi(\theta) \cdot \ell(\theta; D) \quad \forall \theta \in \Theta.$$

14

Bayes' theorem and fuzzy information

Looking at real data and realistic a priori distributions two kinds of *fuzzy information* appear. First, fuzzy samples and secondly, fuzzy a priori distributions. The necessity of generalizing Bayesian inference was pointed out in Viertl (1987).

The description of fuzzy data is given in earlier parts of this book. Fuzzy a priori information can be expressed by fuzzy probability distributions in the sense of Chapter 8.

14.1 Fuzzy a priori distributions

In the case of discrete stochastic model $p(.|\theta); \theta \in \Theta$ and discrete parameter space $\Theta = \{\theta_1, \ldots, \theta_k\}$ fuzzy a priori information concerning the parameter can be expressed by k fuzzy intervals $\pi^*(\theta_1), \ldots, \pi^*(\theta_k)$ with $\pi^*(\theta_j) = \Pr\{\theta_j\}$ for which $\pi^*(\theta_1) \oplus \ldots \oplus \pi^*(\theta_k)$ is a fuzzy interval whose characterizing function $\eta(\cdot)$ fulfills $1 \in C_1 [\eta(\cdot)]$ and the characterizing functions $\xi_j(\cdot)$ of $\pi^*(\theta_j)$ are fulfilling the following:

For each $j \in \{1, \ldots, k\}$ there exists a number

$$p_j \in C_1 \left[\pi^*(\theta_j) \right]$$

such that

$$\sum_{j=1}^{k} p_j = 1.$$

Since $C_\delta \left[\pi^*(\theta_j) \right]$ are closed intervals for all $\delta \in (0; 1]$, the fuzzy probability of a subset $\Theta_1 \subset \Theta$ with $\Theta_1 = \{\theta_{j1}, \dots, \theta_{jm}\}$ is denoted by $P^*(\Theta_1)$. The δ-cuts of $P^*(\Theta_1)$ are defined to be the set of all possible sums of numbers $x_j \in C_\delta(p_{jl}^*), l = 1(1)m$, obeying

$$\sum_{j=1}^{k} x_j = 1.$$

Remark 14.1: From the above definition it follows that $P^*(\emptyset) = 0$ and $P^*(\Theta) = 1$, and the inequalities from Remark 8.2 in Chapter 8.

For continuous parameter space Θ a *fuzzy a priori distribution* $P^*(\cdot)$ on Θ is given by a fuzzy valued density function $\pi^*(\cdot)$ on Θ as described in Section 8.1.

14.2   Updating fuzzy a priori distributions

For stochastic quantities X with parametric probability distribution $p(.|\theta)$ or $f(.|\theta)$, given observed data $\overset{\circ}{x}_1, \cdots, \overset{\circ}{x}_n$ the a priori distribution has to be updated in order to obtain up-to-date information concerning the parameter θ.

This updated information is the conditional distribution of $\widetilde{\theta}$, shown as $P^* \left(\cdot | \overset{\circ}{x}_1, \cdots, \overset{\circ}{x}_n \right)$, conditional on the sample $\overset{\circ}{x}_1, \cdots, \overset{\circ}{x}_n$.

For a discrete stochastic model $X \sim p(\cdot|\theta); \theta \in \Theta$ with discrete parameter space

$$\Theta = \{\theta_1, \dots, \theta_k\}$$

the a posteriori distribution $P^* \left(\cdot | \overset{\circ}{x}_1 \dots, \overset{\circ}{x}_n \right)$ is given by its conditional point probabilities $p^* \left(\theta_j | \overset{\circ}{x}_1 \dots, \overset{\circ}{x}_n \right), j = 1(1)k$. These fuzzy point probabilities are obtained as conditional probabilities

$$
\pi^* \left(\theta_j | \overset{\circ}{x}_1, \dots, \overset{\circ}{x}_n \right) = \frac{\Pr \left(\overset{\circ}{x}_1, \dots, \overset{\circ}{x}_n | \theta_j \right) \pi^*(\theta_j)}{\sum\limits_{j=1}^{k} \Pr \left(\overset{\circ}{x}_1, \dots, \overset{\circ}{x}_n | \theta_j \right) \pi^*(\theta_j)}
$$

$$
= \frac{\left[\prod\limits_{i=1}^{n} p \left(\overset{\circ}{x}_i | \theta_j \right) \right] \pi^*(\theta_j)}{\sum\limits_{j=1}^{k} \left[\prod\limits_{i=1}^{n} p \left(\overset{\circ}{x}_i | \theta_j \right) \right] \pi^*(\theta_j)} \quad \text{for } j = 1(1)k.
$$

This can be written using the likelihood function as

$$\pi^*\left(\theta_j | \overset{\circ}{x}_1, \cdots, \overset{\circ}{x}_n\right) = \frac{\pi^*(\theta_j) \cdot l\left(\theta_j; \overset{\circ}{x}_1, \cdots, \overset{\circ}{x}_n\right)}{\sum\limits_{j=1}^{k} \pi^*(\theta_j) \cdot l\left(\theta_j; \overset{\circ}{x}_1, \ldots, \overset{\circ}{x}_n\right)}.$$

The value $\pi^*(\theta_j) \cdot l\left(\theta_j; \overset{\circ}{x}_1, \ldots, \overset{\circ}{x}_n\right)$ is a product of a fuzzy interval $\pi^*(\theta_j)$ and a real number $l\left(\theta_j; \overset{\circ}{x}_1, \ldots, \overset{\circ}{x}_n\right)$.

Therefore $\pi^*(\theta_j) \cdot l\left(\theta_j; \overset{\circ}{x}_1, \ldots, \overset{\circ}{x}_n\right)$ is also a fuzzy interval.

Using the generalized arithmetic for fuzzy numbers (cf. Chapter 3) the *fuzzy a posteriori probabilities* $\pi^*\left(\theta_j | \overset{\circ}{x}_1, \ldots, \overset{\circ}{x}_n\right)$ can be calculated.

In order to make the notation compact we use the abbreviation $\overset{\circ}{\underline{x}} = \left(\overset{\circ}{x}_1, \ldots, \overset{\circ}{x}_n\right)$.

Using this, the δ-cuts $C_\delta\left(\pi^*(\theta_j | \overset{\circ}{\underline{x}})\right)$ of the a posteriori probabilities are given, using the δ-cuts $C_\delta\left(\pi^*(\theta_j)\right) = \left[\underline{\pi}_\delta(\theta_j); \overline{\pi}_\delta(\theta_j)\right]$ of the a priori probabilities, by

$$\overline{\pi}_\delta(\theta_j | \overset{\circ}{\underline{x}}) = \frac{\overline{\pi}_\delta(\theta_j) \cdot l(\theta_j; \overset{\circ}{\underline{x}})}{\sum\limits_{j=1}^{k} \underline{\pi}_\delta(\theta_j) \cdot l(\theta_j; \overset{\circ}{\underline{x}})}$$

and

$$\underline{\pi}_\delta(\theta_j | \overset{\circ}{\underline{x}}) = \frac{\underline{\pi}_\delta(\theta_j) \cdot l(\theta_j; \overset{\circ}{\underline{x}})}{\sum\limits_{j=1}^{k} \overline{\pi}_\delta(\theta_j) \cdot l(\theta_j; \overset{\circ}{\underline{x}})} \qquad \text{for all } \delta \in (0; 1].$$

Recalling the sequential property of the Bayesian updating procedure, i.e. splitting the data $\overset{\circ}{x}_1, \ldots, \overset{\circ}{x}_n$ into two parts $\overset{\circ}{\underline{x}}_1 = (\overset{\circ}{x}_1, \ldots, \overset{\circ}{x}_m)$ and $\overset{\circ}{\underline{x}}_2 = (\overset{\circ}{x}_{m+1}, \ldots, \overset{\circ}{x}_n)$, and calculating $\pi^*(\cdot | \overset{\circ}{\underline{x}}_1)$, then using this as a priori distribution for the next sample $\overset{\circ}{\underline{x}}_2$, the resulting a posteriori distribution should be the same as if the whole sample $\overset{\circ}{x}_1, \ldots \overset{\circ}{x}_n$ would be used in one step to calculate the a posteriori distribution.

Unfortunately the above generalization of Bayes' theorem fails to keep this. This is explained below for the continuous case $X \sim f(\cdot | \theta); \theta \in \Theta$ with fuzzy a priori density $\pi^*(\cdot)$ on the continuous parameter space Θ.

Considering a sample $(x_1, \ldots, x_n) = \underline{x}$ of X the δ-level curves would be given by

$$\overline{\pi}_\delta(\theta | \underline{x}) = \frac{\overline{\pi}_\delta(\theta) \cdot l(\theta; \underline{x})}{\int\limits_{\Theta} \underline{\pi}_\delta(\theta) \cdot l(\theta; \underline{x}) d\theta}$$

and

$$\underline{\pi}_\delta(\theta|\underline{x}) = \frac{\underline{\pi}_\delta(\theta) \cdot l(\theta;\underline{x})}{\int\limits_\Theta \overline{\pi}_\delta(\theta) \cdot l(\theta;\underline{x})d\theta} \quad \text{for all } \delta \in (0;1].$$

Using the splitting of the sample $x_1,\ldots,x_n$ above into $x_1,\ldots,x_m$ and $x_{m+1},\ldots,x_n$ we obtain

$$\overline{\pi}_\delta(\theta|\underline{x}_1) = \frac{\underline{\pi}_\delta(\theta) \cdot l(\theta;\underline{x}_1)}{\int\limits_\Theta \underline{\pi}_\delta(\theta) \cdot l(\theta;\underline{x}_1)d\theta}$$

and

$$\underline{\pi}_\delta(\theta|\underline{x}_1) = \frac{\underline{\pi}_\delta(\theta) \cdot l(\theta;\underline{x}_1)}{\int\limits_\Theta \overline{\pi}_\delta(\theta) \cdot l(\theta;\underline{x}_1)d\theta}.$$

Using this fuzzy density $\pi^*(\cdot|\underline{x}_1)$ as a priori distribution for the second updating based on the second sample $(x_{m+1},\ldots,x_n) = \underline{x}_2$ we obtain for the lower δ-level curve:

$$\underline{\pi}_\delta(\theta|\underline{x}_1,\underline{x}_2) = \frac{\underline{\pi}_\delta(\theta|\underline{x}_1) \cdot l(\theta;\underline{x}_2)}{\int\limits_\Theta \overline{\pi}_\delta(\theta|\underline{x}_1) \cdot l(\theta;\underline{x}_2)d\theta}$$

$$= \frac{\frac{\underline{\pi}_\delta(\theta)\cdot l(\theta;\underline{x}_1)}{\int\limits_\Theta \overline{\pi}_\delta(\theta)\cdot l(\theta;\underline{x}_1)d\theta} l(\theta;\underline{x}_2)}{\int\limits_\Theta \frac{\overline{\pi}_\delta(\theta)\cdot l(\theta;\underline{x}_1)}{\int\limits_\Theta \underline{\pi}_\delta(\theta;\underline{x}_1)d\theta} l(\theta;\underline{x}_2)d\theta}$$

$$= \frac{\int\limits_\Theta \underline{\pi}_\delta(\theta) \cdot l(\theta;\underline{x}_1)d\theta}{\int\limits_\Theta \overline{\pi}_\delta(\theta) \cdot l(\theta;\underline{x}_1)d\theta} \cdot \frac{\underline{\pi}_\delta(\theta) \cdot l(\theta;\underline{x}_1) \cdot l(\theta;\underline{x}_2)}{\int\limits_\Theta \overline{\pi}_\delta(\theta) \cdot l(\theta;\underline{x}_1) \cdot l(\theta;\underline{x}_2)d\theta}$$

Observing $l(\theta;\underline{x}_1) \cdot l(\theta;\underline{x}_2) = l(\theta;\underline{x})$ and by

$$\frac{\int\limits_\Theta \underline{\pi}_\delta(\theta) \cdot l(\theta;\underline{x}_1)d\theta}{\int\limits_\Theta \overline{\pi}_\delta(\theta) \cdot l(\theta;\underline{x}_1)d\theta} = C \le 1$$

we obtain

$$\underline{\pi}_\delta(\theta|\underline{x}_1,\underline{x}_2) = C \cdot \underline{\pi}_\delta(\theta|\underline{x}).$$

Therefore the results differ by a constant C. In order to keep the sequential nature of the updating procedure in Bayes' theorem it is necessary to use a common denominator for $\pi_\delta(\cdot|\underline{x})$ and $\overline{\pi}_\delta(\cdot|\underline{x})$.

The natural way to do this is to use the mean value

$$\frac{\pi_\delta(\theta) + \overline{\pi}_\delta(\theta)}{2}.$$

Using this, the following form of the Bayes' updating procedure is obtained:

$$\overline{\pi}_\delta(\theta|\underline{x}) := \frac{\overline{\pi}_\delta(\theta) \cdot l(\theta;\underline{x})}{\int_\Theta \frac{\pi_\delta(\theta)+\overline{\pi}_\delta(\theta)}{2} \cdot l(\theta;\underline{x})d\theta}$$

and

$$\underline{\pi}_\delta(\theta|\underline{x}) := \frac{\pi_\delta(\theta) \cdot l(\theta;\underline{x})}{\int_\Theta \frac{\pi_\delta(\theta)+\overline{\pi}_\delta(\theta)}{2} \cdot l(\theta;\underline{x})d\theta}.$$

Using this generalization of Bayes' theorem the two-step calculated fuzzy a posteriori density $\pi^*(\cdot|\underline{x}_1,\underline{x}_2)$ equals the one-step calculated a posteriori density $\pi^*(\cdot|\underline{x})$. In order to prove this we use the abbreviation

$$N(\underline{x}) := \int_\Theta \frac{\pi_\delta(\theta) + \overline{\pi}_\delta(\theta)}{2} \cdot l(\theta;\underline{x})d\theta.$$

For the lower δ-level curve of $\pi^*(\cdot|\underline{x}_1,\underline{x}_2)$ with actual a priori density $\pi^*(\cdot|\underline{x}_1)$ we obtain

$$\underline{\pi}_\delta(\theta|\underline{x}_1,\underline{x}_2) = \frac{\pi_\delta(\theta|\underline{x}_1) \cdot l(\theta;\underline{x}_2)}{\int_\Theta \frac{\pi_\delta(\theta|\underline{x}_1)+\overline{\pi}_\delta(\theta|\underline{x}_1)}{2} \cdot l(\theta;\underline{x}_2)d\theta}$$

$$= \frac{N(\underline{x}_1)^{-1} \cdot \pi_\delta(\theta) \cdot l(\theta;\underline{x}_1) \cdot l(\theta;\underline{x}_2)}{\int_\Theta N(\underline{x}_1)^{-1} \cdot \frac{\pi_\delta(\theta)\cdot l(\theta;\underline{x}_1)+\overline{\pi}_\delta(\theta)\cdot l(\theta;\underline{x}_1)}{2} \cdot l(\theta;\underline{x}_2)d\theta}$$

$$= \frac{\pi_\delta(\theta) \cdot l(\theta;\underline{x}_1) \cdot l(\theta;\underline{x}_2)}{\int_\Theta \frac{\pi_\delta(\theta)+\overline{\pi}_\delta(\theta)}{2} \cdot l(\theta;\underline{x}_1) \cdot l(\theta;\underline{x}_2)d\theta} = \underline{\pi}_\delta(\theta|\underline{x}).$$

The upper δ-level curve $\overline{\pi}_\delta(\cdot|\underline{x}_1,\underline{x}_2)$ is proved to be equal to $\overline{\pi}_\delta(\theta|\underline{x})$ in an analogous way.

So for classical samples $\underline{x} = (x_1,\ldots,x_n)$ a suitable generalization of Bayes' theorem to the case of fuzzy a priori distributions is given.

The situation of fuzzy a priori distribution and fuzzy sample $x_1^*,\ldots,x_n^*$ is solved in Chapter 15.

14.3   Problems

(a) Looking at a fuzzy discrete probability distribution P^* on the finite set $\{x_1, \ldots, x_k\}$ with fuzzy point probabilities $p^*(x_j)$, $j = 1(1)k$ explain the validity of the rules for a fuzzy probability distribution given in Section 8.3.

(b) Prove the validity of the sequential nature for the upper δ-level curves of the fuzzy a posteriori density $\pi^*(\cdot | \underline{x})$.

(c) Show that the above procedure for fuzzy a posteriori densities gives the classical a posteriori density if the a priori density is a classical density.

15

Generalized Bayes' theorem

For continuous stochastic models $X \sim f(\cdot|\theta); \theta \in \Theta$ with continuous parameter space in general a priori distributions as well as observations are fuzzy. Therefore it is necessary to generalize Bayes' theorem to this situation.

15.1 Likelihood function for fuzzy data

In the case of fuzzy data $x_1^*, \ldots, x_n^*$ the likelihood function $l(\theta; x_1, \ldots, x_n)$ has to be generalized to the situation of fuzzy variables $x_1^*, \ldots, x_n^*$. The basis for that is the combined fuzzy sample element $\underline{x}^*$ from Chapter 10. Then the generalized likelihood function $l^*(\theta; \underline{x}^*)$ is represented by its δ-level functions $\underline{l}_\delta(\cdot; \underline{x}^*)$ and $\bar{l}_\delta(\cdot; \underline{x}^*)$ for all $\delta \in (0; 1]$.

For the δ-cuts of the fuzzy value $l^*(\theta; \underline{x}^*)$ we have

$$C_\delta(l^*(\theta; \underline{x}^*)) = \left[\underline{l}_\delta(\theta; \underline{x}^*), \bar{l}_\delta(\theta; \underline{x}^*) \right].$$

Using this and the construction from Chapter 14 in order to keep the sequential property of the updating procedure in Bayes' theorem, the generalization of Bayes' theorem to the situation of fuzzy a priori distribution and fuzzy data is possible.

Remark 15.1: The generalized likelihood function $l^*(\theta; \underline{x}^*)$ is a fuzzy valued function in the sense of Section 3.6, i.e. $l^* : \Theta \to \mathcal{F}_I([0; \infty))$.

15.2 Bayes' theorem for fuzzy a priori distribution and fuzzy data

Using the averaging procedure of δ-level curves of the a priori density from Section 14.2 and combining it with the generalized likelihood function from Section

Statistical Methods for Fuzzy Data   Reinhard Viertl
© 2011 John Wiley & Sons, Ltd

15.1, the generalization of Bayes' theorem is possible. The construction is based on δ-level functions.

Based on a fuzzy a priori density $\pi^*(\cdot)$ on Θ with δ-level functions $\underline{\pi}_\delta(\cdot)$ and $\overline{\pi}_\delta(\cdot)$, and a fuzzy sample $x_1^*, \ldots, x_n^*$ with combined fuzzy sample $\underline{x}^*$ whose vector-characterizing function is $\xi(., \ldots, .)$, the characterizing function $\psi_{l^*(\theta;\underline{x}^*)}(\cdot)$ of $l^*(\theta; \underline{x}^*)$ is obtained by the extension principle from Section 3.1, i.e.

$$\psi_{l^*(\theta;\underline{x}^*)}(y) = \begin{cases} \sup\left\{\xi(\underline{x}) : l(\theta; \underline{x}) = y\right\} \\ 0 \qquad\qquad\qquad\qquad \text{if } \nexists\, \underline{x} : l(\theta; \underline{x}) = y \end{cases} \qquad \forall y \in \mathbb{R}.$$

The δ-level curves of the fuzzy a posteriori density $\pi^*(\cdot|x_1^*, \ldots x_n^*) = \pi^*(\cdot|\underline{x}^*)$ are defined in the following way:

$$\overline{\pi}_\delta(\theta|\underline{x}^*) := \frac{\overline{\pi}_\delta(\theta) \cdot \overline{l}_\delta(\theta; \underline{x}^*)}{\int\limits_\Theta \frac{1}{2}\left[\underline{\pi}_\delta(\theta) \cdot \underline{l}_\delta(\theta; \underline{x}^*) + \overline{\pi}_\delta(\theta) \cdot \overline{l}_\delta(\theta; \underline{x}^*)\right] d\theta}$$

and

$$\underline{\pi}_\delta(\theta|\underline{x}^*) := \frac{\underline{\pi}_\delta(\theta) \cdot \underline{l}_\delta(\theta; \underline{x}^*)}{\int\limits_\Theta \frac{1}{2}\left[\underline{\pi}_\delta(\theta) \cdot \underline{l}_\delta(\theta; \underline{x}^*) + \overline{\pi}_\delta(\theta) \cdot \overline{l}_\delta(\theta; \underline{x}^*)\right] d\theta}$$

for all $\delta \in (0; 1]$.

Remark 15.2: For the definition of the δ-level curves of the fuzzy a posteriori density the sequentially calculated a posteriori density $\pi^*(\cdot|\underline{x}_1^*, \underline{x}_2^*)$ is the same as $\pi^*(\cdot|\underline{x}^*)$, where $x_1^*, \ldots, x_n^* = x_1^*, \ldots, x_m^*, x_{m+1}^*, \ldots, x_n^*$ and $\underline{x}_1^* = (x_1^*, \ldots, x_m^*)$ and $\underline{x}_2^* = (x_{m+1}^*, \ldots, x_n^*)$. To see this we calculate the δ-cut

$$C_\delta\left(\pi^*(\theta|\underline{x}_1^*, \underline{x}_2^*)\right) = \left[\underline{\pi}_\delta(\theta|\underline{x}_1^*, \underline{x}_2^*); \overline{\pi}_\delta(\theta|\underline{x}_1^*, \underline{x}_2^*)\right]$$

using $\pi^*(\cdot|\underline{x}_1^*)$ and $\underline{x}_2^*$, and the abbreviation

$$N(\underline{x}^*) = \frac{1}{2}\left[\int\limits_\Theta \underline{\pi}_\delta(\theta) \cdot \underline{l}_\delta(\theta; \underline{x}^*)d\theta + \int\limits_\Theta \overline{\pi}_\delta(\theta) \cdot \overline{l}_\delta(\theta; \underline{x}^*)d\theta\right]$$

$$= \int\limits_\Theta \frac{1}{2}\left[\underline{\pi}_\delta(\theta) \cdot \underline{l}_\delta(\theta; \underline{x}^*) + \overline{\pi}_\delta(\theta) \cdot \overline{l}_\delta(\theta; \underline{x}^*)d\theta\right].$$

$$\underline{\pi}_\delta(\theta|\underline{x}_1^*, \underline{x}_2^*) = \frac{\underline{\pi}_\delta(\theta|\underline{x}_1^*) \cdot \underline{l}_\delta(\theta; \underline{x}_2^*)}{\int\limits_\Theta \frac{1}{2}\left[\underline{\pi}_\delta(\theta|\underline{x}_1^*) \cdot \underline{l}_\delta(\theta; \underline{x}_2^*) + \overline{\pi}_\delta(\theta|\underline{x}_1^*) \cdot \overline{l}_\delta(\theta; \underline{x}_2^*)\right] d\theta}$$

$$= \frac{N(\underline{x}_1^*)^{-1}\underline{\pi}_\delta(\theta) \cdot \underline{l}_\delta(\theta; \underline{x}_1^*) \cdot \underline{l}(\theta; \underline{x}_2^*)}{\int\limits_\Theta \frac{1}{2}N(\underline{x}_1^*)^{-1}\left[\underline{\pi}_\delta(\theta) \cdot \underline{l}_\delta(\theta; \underline{x}_1^*) \cdot \underline{l}_\delta(\theta; \underline{x}_2^*) + \overline{\pi}_\delta(\theta) \cdot \overline{l}_\delta(\theta; \underline{x}_1^*) \cdot \overline{l}_\delta(\theta; \underline{x}_2^*)\right] d\theta}$$

$$= \frac{\underline{\pi}_\delta(\theta) \cdot \underline{L}_\delta(\theta; \underline{x}_1^*) \cdot \underline{L}_\delta(\theta; \underline{x}_2^*)}{\int_\Theta \frac{1}{2} \left[\underline{\pi}_\delta(\theta) \cdot \underline{L}_\delta(\theta; \underline{x}_1^*) \cdot \underline{L}_\delta(\theta; \underline{x}_2^*) + \overline{\pi}_\delta(\theta) \cdot \overline{l}_\delta(\theta; \underline{x}_1^*) \cdot \overline{l}_\delta(\theta; \underline{x}_2^*) \right] d\theta}$$

$$= \frac{\underline{\pi}_\delta(\theta) \cdot \underline{L}_\delta(\theta; \underline{x}^*)}{\int_\Theta \frac{1}{2} \left[\underline{\pi}_\delta(\theta) \cdot \underline{L}_\delta(\theta; \underline{x}^*) + \overline{\pi}_\delta(\theta) \cdot \overline{l}_\delta(\theta; \underline{x}^*) \right] d\theta} = \underline{\pi}_\delta(\theta | \underline{x}^*).$$

The proof for the upper δ-level curves is analogous and left to the reader as a problem.

Remark 15.3: For standard a priori densities and precise data this concept reduces to the classical Bayes' theorem.

Example 15.1 For $X \sim Ex_\theta; \theta \in \Theta = (0; \infty)$, i.e. exponential distribution with density

$$f(x|\theta) = \frac{1}{\theta} e^{-\frac{x}{\theta}} I_{(0;\infty)}(x)$$

a fuzzy a priori density $\pi^*(\cdot)$ on Θ is given for which some δ-level curves are depicted in Figure 15.1.

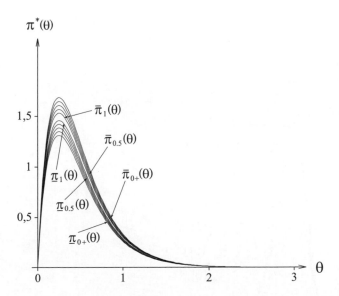

$\underline{\pi}_{0+}(\theta)$ and $\overline{\pi}_{0+}(\theta)$ are the boundary points of $\left\{ x \in \mathbb{R} : \xi_{\pi^*(\theta)}(x) > 0 \right\}$ where $\xi_{\pi^*(\theta)}(\cdot)$ is the characterizing function of the fuzzy interval $\pi^*(\theta)$.

Figure 15.1   Some δ-level curves of $\pi^*(\cdot)$.

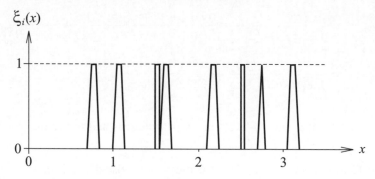

$\xi_i(\cdot)$ is the characterizing function of x_i^*.

Figure 15.2   Fuzzy sample.

In Figure 15.2 the characterizing functions of eight fuzzy observations $x_1^*, \ldots, x_8^*$ of X are given.

Applying the generalized Bayes' theorem the δ-level curves of the a posteriori density $\pi^*(\cdot|x_1^*, \ldots, x_8^*)$ are obtained.

For the present example some δ-level curves $\underline{\pi}_\delta(\cdot|x_1^*, \ldots, x_8^*)$ and $\overline{\pi}_\delta(\cdot|x_1^*, \ldots, x_8^*)$ are depicted in Figure 15.3.

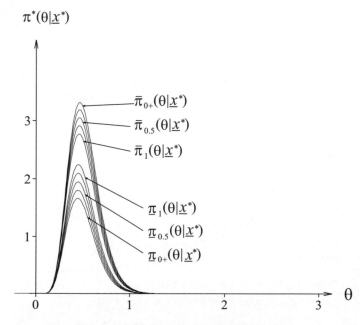

Figure 15.3   Fuzzy a posteriori density.

15.3   Problems

(a) Prove the validity of the sequential updating in the generalized Bayes' theorem for the upper δ-level curves of the fuzzy a posteriori density.
(b) Prove that in case of precise data and classical a priori density the generalized Bayes' theorem coincides with the classical Bayes' theorem.

16

Bayesian confidence regions

In contrast to classical confidence regions in objectivist statistics Bayesian confidence regions have an immediate probabilistic interpretation also after the sample is observed. This is by the availability of the a posteriori distribution $\pi(\cdot|D)$. A Bayesian confidence region for the parameter θ of a stochastic model $X \sim f(\cdot|\theta); \theta \in \Theta$ and confidence level $1 - \alpha$ is a subset $\Theta_{1-\alpha}$ of Θ for which $\Pr\{\tilde{\theta} \in \Theta_{1-\alpha}|D\} = 1 - \alpha$.

In the case of continuous parameter space Θ the confidence region Θ_{1-a} is defined by

$$\int_{\Theta_{1-\alpha}} \pi(\theta|D)d\theta = 1 - \alpha.$$

For fuzzy data D^* the concept of Bayesian confidence regions has to be generalized.

16.1 Bayesian confidence regions based on fuzzy data

In the case of classical a priori distributions the generalization to the situation of fuzzy samples $x_1^*, \ldots, x_n^*$ is possible using the combined fuzzy sample $\underline{x}^*$, whose vector-characterizing is denoted by $\xi(., \ldots, .)$. The generalization is similar to that in Section 12.2.

Let $\Theta_{\underline{x}, 1-\alpha}$ be a standard Bayesian confidence region for θ based on a classical sample $\underline{x} = (x_1, \ldots, x_n)$.

The Bayesian confidence region for θ based on a fuzzy sample, whose combined fuzzy sample has vector-characterizing function $\xi : M_X^n \to [0; 1]$, is a fuzzy subset $\Theta_{1-\alpha}^*$ of Θ whose membership function $\varphi(\cdot)$ is given by its values $\varphi(\theta)$ in the following

Statistical Methods for Fuzzy Data   Reinhard Viertl
© 2011 John Wiley & Sons, Ltd

way:

$$\varphi(\theta) = \begin{cases} \sup\left\{\xi(\underline{x}) : \theta \in \Theta_{\underline{x},1-\alpha}\right\} & \text{if } \exists \underline{x} : \theta \in \Theta_{\underline{x},1-\alpha} \\ 0 & \text{if } \nexists \underline{x} : \theta \in \Theta_{\underline{x},1-\alpha} \end{cases} \quad \forall \theta \in \Theta$$

Remark 16.1: For classical samples $\underline{x} = (x_1, \ldots, x_n)$ the membership function $\varphi(\cdot)$ reduces to the indicator function of $\Theta_{\underline{x},1-\alpha}$, i.e.

$$\varphi(\theta) = I_{\Theta_{\underline{x},1-\alpha}}(\theta) \quad \forall \theta \in \Theta.$$

The proof of this assertion is left to the reader as a problem.

16.2   Fuzzy HPD-regions

Classical HPD-regions $\Theta_{H,1-\alpha}$ are defined for standard a posteriori densities $\pi(\cdot|\underline{x})$ in the following way:

$\Theta_{H,1-\alpha}$ is a subset of Θ for which

$$\int_{\Theta_{H,1-\alpha}} \pi(\theta|\underline{x})\,d\theta = 1 - \alpha \tag{16.1}$$

such that

$$\pi(\theta|\underline{x}) \geq C \quad \forall \theta \in \Theta_{H,1-\alpha} \tag{16.2}$$

where C is the maximal possible constant such that (16.1) is fulfilled.

In the case of classical a priori distribution and fuzzy data $x_1^*, \ldots, x_n^*$ with combined fuzzy sample $\underline{x}^* \hat{=} \xi(., \ldots, .)$, the concept of highest posterior density regions can be generalized in the following way:

For confidence level $1 - \alpha$ and all $\theta \in \Theta$, and $\underline{x} \in \text{supp}(\underline{x}^*) = \text{supp}\,[\xi(., \ldots, .)]$, a classical subset $B(\theta, \underline{x})$ of Θ is defined by

$$B(\theta, \underline{x}) := \left\{\theta^1 \in \Theta : \pi(\theta^1|\underline{x}) \geq \pi(\theta|\underline{x})\right\}.$$

The generalized HPD-region for θ based on the fuzzy sample $\underline{x}^*$ is the fuzzy subset $\Theta_{H,1-\alpha}^*$ whose membership function $\psi(\cdot)$ is determined by is values

$$\psi(\theta) = \begin{cases} \sup\left\{\xi(\underline{x}) : \int_{B(\theta,\underline{x})} \pi(\theta^1|\underline{x})d\theta^1 \leq 1-\alpha\right\} & \text{if } \exists \underline{x} : \int_{B(\theta,\underline{x})} \pi(\theta^1|\underline{x})d\theta^1 \leq 1-\alpha \\ 0 & \text{otherwise} \end{cases} \quad \forall \theta \in \Theta,$$

where $\underline{x} \in \text{supp}\,[\xi(., \ldots, .)]$.

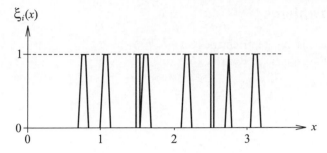

Figure 16.1   Fuzzy sample $x_1^*, \ldots, x_8^*$.

Remark 16.2: For this construction in the case of classical data $\underline{x} = (x_1, \ldots, x_n)$ the resulting membership function $\psi(\cdot)$ is the indicator function of the classical HPD-region, i.e.

$$\psi(\cdot) = I_{\Theta_{H,1-\alpha}}(\cdot).$$

The proof of this is left to the reader as a problem.

Example 16.1 For $X \sim f(x|\theta) = \theta \cdot e^{-\theta x} I_{(0;\infty)}(x)$, $\Theta = (0; \infty)$, and a priori density $\pi(\cdot)$ a Gamma density *Gam*(1, 1), a fuzzy sample $x_1^*, \ldots, x_8^*$, with corresponding characterizing functions $\xi_1(\cdot), \ldots, \xi_8(\cdot)$, is depicted in Figure 16.1. The membership functions of fuzzy HPD-intervals for θ are depicted in Figure 16.2. The different membership functions correspond to different confidence levels $1 - \alpha$.

Remark 16.3: The construction of generalized confidence sets based on fuzzy a posteriori distributions in the general case is an open research topic.

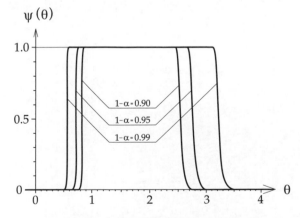

Figure 16.2   Fuzzy HPD-intervals for θ.

16.3   Problems

(a) Prove the statement in Remark 16.1.
(b) Prove the statement in Remark 16.2.

17

Fuzzy predictive distributions

Let $X \sim P_\theta; \theta \in \Theta$ be a stochastic model with a priori distribution $\pi(\cdot)$. Then in standard Bayesian inference the *predictive distribution* of X, given observed data D, is denoted by $X|D$.

There are three practically important situations for predictions:

- discrete X and discrete Θ;

- discrete X and continuous Θ;

- continuous X and continuous Θ.

For the standard situation predictive distributions conditional on observed samples $D = (x_1, \ldots, x_n)$ are available.

In the case of fuzzy a priori distributions and fuzzy samples $(x_1^* \ldots, x_n^*)$ of X the concept of predictive distributions has to be generalized.

17.1 Discrete case

For discrete stochastic quantities $X \sim p(\cdot|\theta), \theta \in \Theta = \{\theta_1, \ldots, \theta_m\}$, and sample $x_1, \ldots, x_n$ of X, the a posteriori probabilities $\pi(\theta_j|x_1, \ldots, x_n) = \Pr(\tilde{\theta} = \theta_j|x_1, \ldots, x_n)$ are obtained as explained at the beginning of Part V.

The predictive distribution of $X|x_1, \ldots, x_n$ is given by its predictive probabilities $p(x|x_1, \ldots, x_n) = \sum_{\theta_j \in \Theta} p(x|\theta_j)\pi(\theta_j|x_1, \ldots x_n)$, for all $x \in M_x$ where M_x is the observation space of X, i.e. the set of all possible values of X.

Statistical Methods for Fuzzy Data   Reinhard Viertl
© 2011 John Wiley & Sons, Ltd

In the case of fuzzy a posteriori distribution $\pi^*(\theta_j|x_1, \ldots, x_n)$ the so-called *fuzzy predictive distribution* of $X|x_1, \ldots, x_n$ is given by its fuzzy probabilities

$$p^*(x|x_1, \ldots, x_n) = \sum_{\theta_j \in \Theta} p(x|\theta_j) \odot \pi^*(\theta_j|x_1, \ldots, x_n).$$

The fuzzy a posteriori probabilities $\pi^*(\theta_j|x_1, \ldots x_n)$ are obtained as explained in Section 14.2.

Remark 17.1: Observations from discrete stochastic quantities can be exact values x_i, whereas observations from continuous quantities are usually fuzzy.

17.2   Discrete models with continuous parameter space

For discrete model $X \sim p(\cdot|\theta); \theta \in \Theta$ where Θ is continuous with a priori density $\pi(\cdot)$ on Θ, and data D, Bayes' theorem has the form

$$\pi(\theta|D) \propto \pi(\theta) \cdot l(\theta; D)$$

where $l(.; D)$ is the likelihood function and $\propto$ stands for 'proportional' which means equal up to a multiplicative constant, i.e.

$$\pi(\theta|D) = \frac{\pi(\theta) \cdot l(\theta; D)}{\int\limits_{\Theta} \pi(\theta) \cdot l(\theta; D)d\theta} \quad \forall \theta \in \Theta.$$

Example 17.1 Let $X \sim A_\theta; \theta \in [0; 1]$ have a Bernoulli distribution with point probabilities $p(1|\theta) = \theta$ and $p(0|\theta) = 1 - \theta$. For a priori density $\pi(\cdot) = I_{[0;1]}(\cdot)$ on $\Theta = [0; 1]$ and observed sample $x_1, \ldots, x_n$ with $x_i \in \{0, 1\}$ the a posteriori density is obtained using the likelihood function

$$l(\theta; x_1, \ldots, x_n) = \prod_{i=1}^{n} \theta^{x_i}(1 - \theta)^{1-x_i} \begin{cases} x_i \in \{0, 1\} \\ \theta \in [0; 1]. \end{cases}$$

Therefore the a posteriori density $\pi(\cdot|x_1, \ldots, x_n)$ is fulfilling

$$\pi(\theta|x_1, \ldots, x_n) \propto \pi(\theta) \cdot l(\theta; x_1, \ldots, x_n) \quad \forall \theta \in [0; 1]$$

or

$$\pi(\theta|x_1, \ldots, x_n) = \frac{\pi(\theta) \cdot \prod_{i=1}^{n} \theta^{x_i}(1 - \theta)^{1-x_i}}{\int\limits_{0}^{1} \pi(\theta) \cdot \prod_{i=1}^{n} \theta^{x_i}(1 - \theta)^{1-x_i} d\theta}.$$

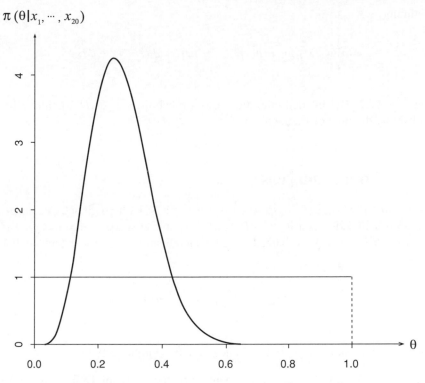

$\pi\left(\theta|x_1, \cdots, x_{20}\right)$

Figure 17.1   A priori and a posteriori density of $\tilde{\theta}$.

By $\pi(\cdot) = I_{[0;1]}(\cdot)$ we obtain

$$\pi(\theta|x_1, \ldots, x_n) \propto \theta^{\sum\limits_{i=1}^{n} x_i} (1 - \theta)^{n - \sum\limits_{i=1}^{n} x_i} I_{[0;1]}(\theta)$$

which is proportional to the density of a beta distribution.

$$\mathrm{Be}\left(\sum_{i=1}^{n} x_i + 1, n - \sum_{i=1}^{n} x_i + 1\right).$$

In Figure 17.1 the uniform a priori density on [0;1] and a corresponding a posteriori density is depicted. From this the information gain by the reduced uncertainty concerning θ can be seen.

The predictive distribution of X is given by the point probabilities $p(x|D) = \int_{\Theta} p(x|\theta) \cdot \pi(\theta|D)d\theta$ in the case of continuous parameter space Θ.

For fuzzy a posteriori density $\pi^*(\cdot)$ of $\tilde{\theta}$, the generalized point probabilities $p^*(x|D^*)$ are obtained by application of the generalized integration from Section 3.6,

Definition 3.3:

$$p^*(x|D^*) := \int_{\Theta} p(x|\theta) \odot \pi^*(\theta|D^*)d\theta, \quad \forall x \in M_X.$$

Remark 17.2: By this definition the fuzzy probabilities $p^*(x|D^*)$ define a fuzzy probability distribution on the observation space M_X of X.

17.3   Continuous case

For continuous stochastic models $X \sim f(\cdot|\theta)$, $\theta \in \Theta$ with continuous parameter θ and fuzzy a posteriori density $\pi^*(\cdot|D^*)$ on the parameter space Θ, the *fuzzy predictive density* $f^*(\cdot|D^*)$ of X is defined by the generalization of the standard predictive density:

$$f^*(x|D^*) := \int_{\Theta} f(x|\theta) \odot \pi^*(\theta|D^*)d\theta$$

where the integral is the generalized integral defined in Section 3.6.

Remark 17.3: $f^*(\cdot|D^*)$ is a fuzzy density in the sense of Definition 8.1 from Section 8.1.

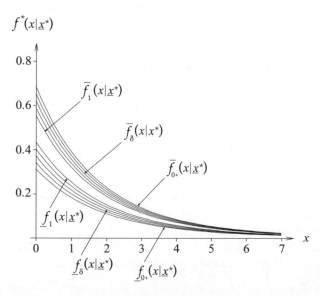

Figure 17.2   δ-level curves of $f^*(\cdot|\underline{x}^*)$.

Example 17.2 In continuation of Example 16.1 the calculation of the fuzzy predictive density $f^*(\cdot|\underline{x}^*)$ can be calculated.

In Figure 17.2 some δ-level curves of $f^*(\cdot|\underline{x}^*)$ are depicted for the fuzzy sample $x_1^*, \ldots, x_8^*$ from Figure 16.1. The combined fuzzy sample $\underline{x}^*$ is constructed by the minimum combination rule.

17.4   Problems

(a) Prove the assertion in Remark 17.2.
(b) Prove the assertion in Remark 17.3.

18

Bayesian decisions and fuzzy information

Decisions are frequently connected with utility or loss. The quantification of the utility $U(d)$ of a decision d can be a difficult task. Therefore a suitable model is to describe the utility by a fuzzy number $U^*(d)$. Moreover the utility of a decision depends on the state of the considered system. Therefore the utility depends on both the state θ and the decision d, i.e. $U(\theta, d)$.

18.1 Bayesian decisions

In Bayesian decision analysis it is possible to include a priori information in the form of a priori distributions, and also a posteriori distributions on the space Θ of possible states of the considered decision situation.

Let $\Theta = \{\theta: \theta$ possible state$\}$ be the state space, $\mathcal{D} = \{d : d$ possible decision$\}$ be the set of possible decisions, and P be a probability distribution on the state space Θ. Moreover let $U(.,.)$ be a utility function, i.e.

$$U: \Theta \times \mathcal{D} \to [0; \infty),$$

where $U(\theta, d)$ denotes the utility of the decision d if the system is in state θ.

Assuming that the corresponding sums or integrals exist, the basis for optimal decisions is the *expected utility*

$$\mathbb{E}_P U(\tilde{\theta}, d).$$

Here $\tilde{\theta}$ is the stochastic quantity describing the random character of the system.

Statistical Methods for Fuzzy Data   Reinhard Viertl
© 2011 John Wiley & Sons, Ltd

Definition 18.1: Based on the context above the Bayesian decision d_B is the solution of the following optimization problem:

$$\mathbb{E}_P U(\tilde{\theta}, d_B) = \max \left\{ \mathbb{E}_P U(\tilde{\theta}, d) : d \in \mathcal{D} \right\}.$$

Therefore the Bayesian decision d_B is maximizing the expected utility.

Remark 18.1: If the approach by loss considerations is used, instead of utility functions so-called *loss functions* $L(\theta, d)$ are considered. In this case Bayesian decisions are defined by minimizing expected loss $\mathbb{E}_P L(\tilde{\theta}, d)$.

Remark 18.2: Expected utility and expected loss are calculated using the probability distribution P in the following way:

$$\mathbb{E}_P U(\tilde{\theta}, d) = \int_{\Theta} U(\theta, d) \, dP(\theta)$$

and

$$\mathbb{E}_P L(\tilde{\theta}, d) = \int_{\Theta} L(\theta, d) \, dP(\theta),$$

respectively.

18.2   Fuzzy utility

In reality it is frequently not possible to find exact utility values. Therefore a more realistic approach to describe utilities is to use *fuzzy utility values* $U^*(\theta, d)$ which are fuzzy intervals. Moreover, in general the uncertainty concerning the state of the system can be fuzzy too. This can be expressed by a fuzzy probability distribution P^* on Θ, for example a fuzzy a posteriori distribution $\pi^*(\cdot | D^*)$.

In this situation, in order to find Bayesian decisions generalized expected utilities have to be calculated, i.e.

$$\mathbb{E}_{P_*} U^*(\tilde{\theta}, d).$$

For classical probability distributions P on Θ the generalized expected utility can be calculated using the generalized integral from Section 3.6:

$$\mathbb{E}_P U^*(\tilde{\theta}, d) = \fint_{\Theta} U^*(\theta, d) \, dP(\theta)$$

is calculated using the δ-level functions

$$\underline{U}_\delta(\cdot, d) \text{ and } \overline{U}_\delta(\cdot, d) \text{ of } U^*(\cdot, d).$$

Remark 18.3: The calculation of the above generalized Stieltjes integral is explained separately for discrete Θ and continuous Θ in the next two sections.

18.3   Discrete state space

Let $\Theta = \{\theta_1, \ldots, \theta_m\}$ be discrete, and P a discrete probability distribution on Θ with point probabilities $p_j = P(\theta_j)$ for $j = 1(1)m$. Then the expected utility of a decision d for the fuzzy utility function $U^*(., d)$ is calculated using δ-level functions in the following way: The δ-cuts of the generalized expected utility $\mathbb{E}_P U^*(\tilde{\theta}, d)$,

$$C_\delta \left[\mathbb{E}_P U^*(\tilde{\theta}, d) \right] = \left[\underline{\mathbb{E}}_\delta U^*(\tilde{\theta}, d); \overline{\mathbb{E}}_\delta U^*(\tilde{\theta}, d) \right],$$

are calculated by

$$\overline{\mathbb{E}}_\delta U^*(\tilde{\theta}, d) = \sum_{j=1}^m \overline{U}_\delta(\theta_j, d) \cdot p_j$$

and

$$\underline{\mathbb{E}}_\delta U^*(\tilde{\theta}, d) = \sum_{j=1}^m \underline{U}_\delta(\theta_j, \ d) \cdot p_j$$

for all $\delta \in (0; 1]$.

The characterizing function $\chi(\cdot)$ of $\mathbb{E}_P U^*(\tilde{\theta}, d)$ is obtained from the construction lemma:

$$\chi(x) = \sup \left\{ \delta \cdot I_{\left[\underline{\mathbb{E}}_\delta U^*(\tilde{\theta}, d); \overline{\mathbb{E}}_\delta U^*(\tilde{\theta}, d) \right]}(x) : \delta \in [0; 1] \right\} \quad \forall x \in \mathbb{R}$$

with $C_0 [\mathbb{E}_P U^*(\tilde{\theta}, d)] = \mathbb{R}$.

In the case of *fuzzy probability distribution* P^* on Θ the generalized expected utility $\mathbb{E}_{P_*} U^*(\tilde{\theta}, d)$ is calculated in the following way:

Let $p_j^* = P^*(\theta_j)$, $j = 1(1)m$ be the fuzzy point probabilities (cf. Section 14.1). The fuzzy expected utility is constructed using the δ-level functions of the utility function and the δ-cuts $C_\delta[p_j^*]$ of p_j^*,

$$C_\delta \left[p_j^* \right] = \left[\underline{p}_{j,\delta}; \overline{p}_{j,\delta} \right] \quad \forall \delta \in (0; 1].$$

$$C_\delta \left[\mathbb{E}_{P^*} U^*(\tilde{\theta}, d) \right] = \left[\underline{\mathbb{E}}_\delta U^*(\tilde{\theta}, d); \overline{\mathbb{E}}_\delta U^*(\tilde{\theta}, d) \right]$$

with

$$\overline{\mathbb{E}}_\delta U^*(\tilde{\theta}, d) = \sum_{j=1}^m \overline{U}_\delta(\theta_j, d) \cdot \overline{p}_{j,\delta}$$

and

$$\underline{\mathbb{E}}_\delta U^*(\tilde{\theta}, d) = \sum_{j=1}^m \underline{U}_\delta(\theta_j, d) \cdot \underline{p}_{j,\delta}$$

$$\forall \delta \in (0; 1].$$

From the δ-cuts the characterizing function $\chi(\cdot)$ of the fuzzy expected utility is obtained by the construction lemma as above.

18.4   Continuous state space

For continuous state space the fuzzy probability distribution P^* on Θ is given by a fuzzy probability density f^* on Θ (cf. Definition 8.1 in Section 8.1). In general for fuzzy utility $U^*(., d)$ with δ-level functions $\underline{U}_\delta(\cdot, d)$ and $\overline{U}_\delta(\cdot, d)$, the fuzzy expected utility

$$\mathbb{E}_{P^*} U^*(\tilde{\theta}, d) = \int_\Theta U^*(\theta, d) \odot f^*(\theta) d\theta$$

is calculated in the following way: Denoting the δ-cuts of $\mathbb{E}_{P^*} U^*(\theta, d)$ by

$$\left[\underline{\mathbb{E}}_\delta U^*(\tilde{\theta}, d); \overline{\mathbb{E}}_\delta U^*(\tilde{\theta}, d)\right],$$

the endpoints are obtained by

$$\overline{\mathbb{E}}_\delta U^*(\tilde{\theta}, d) = \int_\Theta \overline{U}_\delta(\theta, d) \overline{f}_\delta(\theta) d\theta$$

and

$$\underline{\mathbb{E}}_\delta U^*(\tilde{\theta}, d) = \int_\Theta \underline{U}_\delta(\theta, d) \underline{f}_\delta(\theta) d\theta$$

$$\forall \delta \in (0; 1]$$

By the construction lemma the characterizing function of $\mathbb{E}_{P^*} U^*(\tilde{\theta}, d)$ is obtained.

Remark 18.4: In general the resulting generalized expected utility is a fuzzy interval. Contrary to the classical situation, where expected utilities are real numbers, it is not always simple to come to an optimal decision. However, the characterizing functions of fuzzy expected utilities provide valuable information concerning the suitability of a decision.

18.5   Problems

(a) Specialize the described approach to the classical situation of Bayesian decisions. What is the characterizing function of the expected utility in this case?

(b) Under which conditions for fuzzy expected utilities is a Bayesian decision uniquely determined? (*Hint*: Look at the support of the fuzzy expected utility.)

Part VI

REGRESSION ANALYSIS AND FUZZY INFORMATION

In regression analysis usually it is assumed that observed data values are real numbers or vectors. In applications where continuous variables are observed this assumption is unrealistic. Therefore data have to be considered as fuzzy. Another kind of fuzzy information is present in Bayesian regression models. Here the a priori distributions of the parameters in regression models are typical examples of fuzzy information in the sense of fuzzy a priori distributions.

Both kinds of fuzzy information, fuzzy data as well as fuzzy probability distributions for the quantification of a priori knowledge concerning parameters has to be taken into account. This is the subject of this part.

19

Classical regression analysis

Regression models describe nondeterministic dependencies between so-called independent variables x and the corresponding so-called dependent variables y. In standard statistics the values of both kinds of variables are assumed to be numbers or vectors. In reality observed values are not precise numbers but more or less fuzzy. Therefore a generalization of regression analysis to the situation of fuzzy data is necessary. In this chapter regression models are first described, and then the generalization to fuzzy data is explained.

19.1 Regression models

Standard regression models consider k independent deterministic variables $x_1, \cdots, x_k$ and a dependent variable y which is modeled by a stochastic quantity Y whose distribution depends on the values $x_1, \cdots, x_k$ of the independent variables.

Using the notation $x = (x_1, \cdots, x_k)$ the independent variables of the corresponding stochastic quantity describing the dependent variable is denoted by Y_x.

A so-called *parametric regression model* models the expectation $\mathbb{E}Y_x$ of Y_x as a function of the independent variables and the parameters $\theta_0, \theta_1, \cdots, \theta_k$ in the following way:

$$\mathbb{E}Y_x = \psi(x_1, \cdots, x_k; \theta_0, \theta_1, \cdots, \theta_k) \tag{19.1}$$

where $\psi(\cdot)$ is a known function, and $\theta_0, \theta_1, \cdots, \theta_k$ are unknown constants called regression parameters.

Usually the variance of Y_x are assumed to be the same for all $x \in \mathbb{R}^k$:

$$\text{Var } Y_x = \sigma^2 \quad \text{for all} \quad x \tag{19.2}$$

Statistical Methods for Fuzzy Data   Reinhard Viertl
© 2011 John Wiley & Sons, Ltd

Condition (19.2) is called *variance homogeneity* or *homoscedasticity*.

The values of the dependent variable y are considered to be the sum of a deterministic function and a stochastic quantity U with $\mathbb{E}U = 0$ and Var $U = \sigma^2$:

$$Y_x = \psi(x; \boldsymbol{\theta}) + U \tag{19.3}$$

where $\boldsymbol{\theta} = (\theta_0, \theta_1, \cdots, \theta_k)$ denotes the vector of unknown parameters.

Based on observed data $(x_{i1}, \cdots, x_{ik}; y_i)$ for $i = 1(1)n$ the parameters $\theta_0, \theta_1, \cdots, \theta_k$ are estimated using the principle of *least sum of squared distances* (LSS):

$$\sum_{i=1}^{n} [y_i - \psi(x_{i1}, \cdots, x_{ik}; \theta_0, \theta_1, \cdots, \theta_k)]^2 \longrightarrow \min.$$

An estimate of the variance σ^2 is obtained as a function of the number n of data 'points' and

$$\sum_{i=1}^{n} [y_i - \psi(x_{i1}, \cdots, x_{ik}; \theta_0, \theta_1, \cdots, \theta_k)]^2.$$

A specialized type of regression models are *linear regression models* for which Equation (19.1) specializes, using the notation

$$\mathring{x} = (1, x_1, \cdots, x_k) \in \mathbb{R}^{k+1},$$

to be of the following form:

$$\mathbb{E}Y_x = \boldsymbol{\theta}\,\mathring{x}^T = (\theta, \theta_1, \cdots, \theta_k) \begin{pmatrix} 1 \\ x_1 \\ \vdots \\ x_k \end{pmatrix} = \theta_0 + \sum_{j=1}^{k} \theta_j x_j \tag{19.4}$$

where $\boldsymbol{\theta} \in \mathbb{R}^{k+1}$.

For linear regression models under certain conditions the solution $\widehat{\theta}_0, \widehat{\theta}_1, \cdots, \widehat{\theta}_k$ of the LSS problem can be given explicitly. This is stated in the well known Gauss–Markov theorem.

In order to formulate the Gauss–Markov theorem the so-called *data matrix* $\mathcal{X}$ is useful.

For data $(x_{i1}, \cdots, x_{ik}; y_i)$, $i = 1(1)n$ the data matrix is defined by

$$\mathcal{X} = \begin{pmatrix} 1 & x_{11} & x_{12} & \cdots & x_{1k} \\ 1 & x_{21} & x_{22} & \cdots & x_{2k} \\ \vdots & & & & \\ 1 & x_{n1} & x_{n2} & \cdots & x_{nk} \end{pmatrix} \qquad \text{with} \quad x_{ij} \in \mathbb{R}.$$

The vector of dependent values y is defined by

$$y = \begin{pmatrix} y_1 \\ y_2 \\ \vdots \\ y_n \end{pmatrix} \qquad \text{with} \quad y_i \in \mathbb{R}.$$

The vector x_i is defined by

$$x_i = (x_{i1}, \cdots, x_{ik}) \quad \text{for} \quad i = 1(1)n.$$

So the data matrix is given by

$$\mathcal{X} = \begin{pmatrix} \mathring{x}_1 \\ \mathring{x}_2 \\ \vdots \\ \mathring{x}_n \end{pmatrix}.$$

Definition 19.1: Estimators $\widehat{\theta}_j$ for parameters θ_j in linear regression models are called *linear* if

$$\widehat{\theta}_j = \sum_{i=1}^{n} c_{ij} y_i \qquad (19.5)$$

where the coefficients c_{ij} are depending only on the known values x_{ij} of the data matrix $\mathcal{X}$.

Remark 19.1: Using the notation above and

$$C = (c_{ij}) \begin{matrix} i = 1(1)n \\ j = 1(1)k \end{matrix} \qquad (19.6)$$

the vector of estimates $\widehat{\theta} = (\widehat{\theta}_0, \widehat{\theta}_1, \cdots, \widehat{\theta}_k)$ can be expressed in the following way:

$$\widehat{\theta} = C y \qquad (19.7)$$

The following Gauss–Markoff theorem gives the explicit solution of the LSS estimation of $\boldsymbol{\theta}$. Moreover it gives an unbiased estimation for σ^2.

Theorem 19.1: For a linear regression model (19.4) and data $(x_{i1}, \cdots, x_{ik}; y_i)$, $i = 1(1)n$ obeying the following conditions

(1) $n \geq k + 1$;

(2) the elements x_{ij} of the data matrix are known and rank $(\mathcal{X}) = k + 1$;

(3) the stochastic quantities $Y_1, \cdots, Y_n$ denoting the dependent variables Y_{x_i}, $i = 1(1)n$ are pairwise uncorrelated;

(4) Var $Y_i = \sigma^2$ for $i = 1(1)n$, i.e. all variances are identical;

(5) the space of possible parameter values $\boldsymbol{\theta}$ is not a hyperplane in $\mathbb{R}^{k+1}$;

the following holds true:

1) The LSS estimates for the parameters are given by

$$\widehat{\boldsymbol{\theta}}^T = \left(\mathcal{X}^T \mathcal{X}\right)^{-1} \mathcal{X}^T \boldsymbol{Y}$$

where $\mathcal{X}^T$ denotes the transposed matrix of $\mathcal{X}$, and $\boldsymbol{Y}$ is the vector of dependent variables

$$\boldsymbol{Y} = \begin{pmatrix} Y_1 \\ \vdots \\ Y_n \end{pmatrix}.$$

2) The estimators $\widehat{\theta}_j$ above are also the minimum variance unbiased linear estimators (MVULE) for θ_j, $j = 0(1)k$.

3) The variance–covariance matrix of $\widehat{\boldsymbol{\theta}}$, denoted by

$$\mathrm{VCov}(\widehat{\theta}_0, \widehat{\theta}_1, \cdots, \widehat{\theta}_k) = \left(\mathcal{X}^T \mathcal{X}\right)^{-1} \sigma^2.$$

4) An unbiased estimator for the variance σ^2 is given by

$$S^2 = \frac{\left(\boldsymbol{Y} - \mathcal{X}\widehat{\boldsymbol{\theta}}^T\right)^T \left(\boldsymbol{Y} - \mathcal{X}\widehat{\boldsymbol{\theta}}^T\right)}{n - k - 1} = \frac{\sum\limits_{i=1}^{n} \left[Y_i - \widehat{\theta}_0 - \sum\limits_{j=1}^{k} x_{ij}\widehat{\theta}_j \right]^2}{n - k - 1}.$$

The proof is given in books on regression analysis.

For linear regression models $Y_x = \overset{\ast}{x}\,\boldsymbol{\theta} + U$ and parameter estimates

$$\widehat{\boldsymbol{\theta}} = \left(\widehat{\theta}_0, \widehat{\theta}_1, \cdots, \widehat{\theta}_k\right)$$

a prediction for the dependent variable y to independent variables $\boldsymbol{x} = (x_1, \cdots, x_n)$ is given by

$$\widehat{Y}_x = \widehat{\boldsymbol{\theta}}\mathring{\boldsymbol{x}}^T = \mathring{\boldsymbol{x}}\widehat{\boldsymbol{\theta}}^T. \tag{19.8}$$

Proposition 19.1: For the prediction (19.8) the following holds true:

(1) $\widehat{Y}_x$ is an unbiased estimator for $\mathbb{E}Y_x$.

(2) $\text{Var } \widehat{Y}_x = \mathring{\boldsymbol{x}} \left(\mathcal{X}^T\mathcal{X}\right)^{-1} \mathring{\boldsymbol{x}}^T \sigma^2$.

Proof:

(1) $\mathbb{E}\widehat{Y}_x = \mathbb{E}\left(\widehat{\boldsymbol{\theta}}\,\mathring{\boldsymbol{x}}^T\right) = \mathbb{E}\left(\widehat{\theta}_0 + \sum_{j=1}^{k} x_j\widehat{\theta}_j\right) = \mathbb{E}\widehat{\theta}_0 + \sum_{j=1}^{k} x_j\mathbb{E}\widehat{\theta}_j$. By the unbiasedness of $\widehat{\theta}_j$ for θ_j, $j = 0(1)k$, the last expression equals $\theta_0 + \sum_{j=1}^{k} x_j\theta_j = \mathbb{E}Y_x$ by the linear regression model (19.4).

(2) $\text{Var } \widehat{Y}_x = \text{Var}\left(\mathring{\boldsymbol{x}}\,\widehat{\boldsymbol{\theta}}^T\right) = \text{Var}\left(\mathring{\boldsymbol{x}}\left(\mathcal{X}^T\mathcal{X}\right)^{-1}\mathcal{X}^T\boldsymbol{Y}\right)$. $\mathring{\boldsymbol{x}}\left(\mathcal{X}^T\mathcal{X}\right)^{-1}\mathcal{X}^T = z = (z_1, \cdots, z_n)$ is a vector of constants and therefore we obtain $\text{Var}\left(\mathring{\boldsymbol{x}}\left(\mathcal{X}^T\mathcal{X}\right)^{-1}\mathcal{X}^T\boldsymbol{Y}\right) = \text{Var}\left(z\,\boldsymbol{Y}\right) = \text{Var}\left(\sum_{i=1}^{n} z_i Y_i\right)$. Because $Y_1, \cdots, Y_n$ are pairwise uncorrelated we obtain $\text{Var}\left(\sum_{i=1}^{n} z_i Y_i\right) = \sum_{i=1}^{n} z_i^2 \text{Var } Y_i = z\,z^T\sigma^2$.

By $z\,z^T = \left(\mathring{\boldsymbol{x}}\left(\mathcal{X}^T\mathcal{X}\right)^{-1}\mathcal{X}^T\right)\left(\mathring{\boldsymbol{x}}\left(\mathcal{X}^T\mathcal{X}\right)^{-1}\mathcal{X}^T\right)^T = \mathring{\boldsymbol{x}}\left(\mathcal{X}^T\mathcal{X}\right)^{-1}\mathcal{X}^T\mathcal{X}$ $\left[\left(\mathcal{X}^T\mathcal{X}\right)^{-1}\right]^T\mathring{\boldsymbol{x}}^T = \mathring{\boldsymbol{x}}\left[\left(\mathcal{X}^T\mathcal{X}\right)^{-1}\right]^T\mathring{\boldsymbol{x}}^T = \mathring{\boldsymbol{x}}\left(\mathcal{X}^T\mathcal{X}\right)^{-1}\mathring{\boldsymbol{x}}^T$, and we obtain for the variance $\mathring{\boldsymbol{x}}\left(\mathcal{X}^T\mathcal{X}\right)^{-1}\mathring{\boldsymbol{x}}^T\sigma^2$.

19.2 Linear regression models with Gaussian dependent variables

Assuming Gaussian distributions for the dependent variables $Y_i = Y_{x_i}$ the estimators from the Gauss–Markov theorem have additional properties. These can be used to construct confidence intervals for the parameters. In order to obtain the confidence intervals some results are necessary. These are given in the following theorems.

Theorem 19.2: Assuming a linear regression model (19.4) and all conditions from Theorem 19.1 and additionally the stochastic quantities $Y_1, \cdots, Y_n$ to have Gaussian distributions, i.e. $\boldsymbol{Y} = \begin{pmatrix} Y_1 \\ \vdots \\ Y_n \end{pmatrix}$ has multivariate Gaussian distribution $N\left(\mathcal{X}\boldsymbol{\theta}^T, \sigma^2 I_n\right)$

where I_n is the $n \times n$ identity matrix

$$I_n = \begin{pmatrix} 1 & 0 & \cdots & & 0 \\ 0 & 1 & & & \vdots \\ \vdots & & \ddots & 1 & 0 \\ 0 & & \cdots & 0 & 1 \end{pmatrix},$$

then the following holds:

(1) The estimator $\widehat{\theta}$ is also the maximum likelihood estimator for θ.

(2) The distribution of the estimator $\widehat{\theta}$ is a multivariate normal distribution

$$N\left(\begin{pmatrix} \theta_0 \\ \theta_1 \\ \vdots \\ \theta_k \end{pmatrix} ; \left(X^T X \right)^{-1} \sigma^2 \right).$$

(3) The pivotal quantity $\dfrac{(Y - X\widehat{\theta})^T (Y - X\widehat{\theta}^T)}{\sigma^2}$ has chi-square distribution with $n - k - 1$ degrees of freedom, and the pivotal quantity $\dfrac{(\widehat{\theta} - \theta) X^T X (\widehat{\theta} - \theta)^T}{\sigma^2}$ has chi-square distribution with $k + 1$ degrees of freedom.

(4) $\dfrac{\left(Y - X\widehat{\theta}^T \right)^T \left(Y - X\widehat{\theta}^T \right)}{\sigma^2}$ and $\dfrac{(\widehat{\theta} - \theta) X^T X (\widehat{\theta} - \theta)^T}{\sigma^2}$ are statistically independent.

The proof can be found in books on regression analysis.

Proposition 19.2: Under the assumptions of Theorem 19.2 the stochastic vector

$$Y - X\widehat{\theta}^T = \begin{pmatrix} Y_1 & - & \widehat{\theta}_0 & - & \sum_{j=1}^{k} x_{1j}\widehat{\theta}_j \\ & \vdots & & & \\ Y_n & - & \widehat{\theta}_0 & - & \sum_{j=1}^{k} x_{nj}\widehat{\theta}_j \end{pmatrix}$$

is normally distributed and statistically independent from $\widehat{\theta}$.

For the proof compare books on regression analysis.

Now confidence intervals for $\theta_0, \theta_1, \cdots, \theta_k$ and σ^2 can be obtained. These are given in the next theorem.

Theorem 19.3: Under the assumptions of Theorem 19.2 and denoting the elements of $(X^T X)^{-1}$ by s_{ij} the following confidence intervals for θ_j, $j = 0(1)k$, and for σ^2 are obtained:

(1) A confidence interval for θ_j with coverage probability $1 - \alpha$ is given by

$$\left[\widehat{\theta}_j - t_{n-k-1;1-\frac{\alpha}{2}} \cdot \sqrt{s_{jj}} \cdot S; \ \widehat{\theta}_j + t_{n-k-1\,;1-\frac{\alpha}{2}} \cdot \sqrt{s_{jj}} \cdot S\right]$$

where $t_{n-k-1\,;1-\frac{\alpha}{2}}$ is the $\left(1 - \frac{\alpha}{2}\right)$ quantile of the student-distribution with $n - k - 1$ degrees of freedom, and S is the positive square root of S^2 from Theorem 19.1.

(2) A confidence interval for σ^2 with coverage probability $1 - \alpha$ is given by

$$\left[\frac{(n - k - 1)S^2}{\chi^2_{n-k-1;1-\frac{\alpha}{2}}} ; \ \frac{(n - k - 1)S^2}{\chi^2_{n-k-1;\frac{\alpha}{2}}}\right]$$

where $\chi^2_{n-k-1;\ p}$ is the p-quantile of the chi-square distribution with $n - k - 1$ degrees of freedom, S is as above.

Proof:

(1) From Theorem 19.2 we obtain by standardization that $\frac{\widehat{\theta}_j - \theta_j}{\sqrt{s_{jj} \cdot \sigma^2}}$ has standard normal distribution. The pivotal quantity $\frac{(n-k-1)S^2}{\sigma^2}$ is chi-square distributed with $n - k - 1$ degrees of freedom. By Proposition 19.2 the two pivotal quantities are statistically independent. Therefore the stochastic quantity

$$\frac{\left(\widehat{\theta}_j - \theta_j\right) / \sqrt{s_{jj}} \cdot \sigma}{\sqrt{\frac{(n-k-1)S^2}{\sigma^2/(n-k-1)}}} = \frac{\widehat{\theta}_j - \theta_j}{S\sqrt{s_{jj}}}$$

is student distributed with $n - k - 1$ degrees of freedom. Therefore

$$Pr\left\{t_{n-k-1;\frac{\alpha}{2}} \le \frac{\widehat{\theta}_j - \theta_j}{S\sqrt{s_{jj}}} \le t_{n-k-1;1-\frac{\alpha}{2}}\right\} = 1 - \alpha,$$

and identical transformation of the inequalities yields

$$Pr\left\{\widehat{\theta}_j - t_{n-k-1;1-\frac{\alpha}{2}} \cdot \sqrt{s_{jj}} \cdot S \le \theta_j \le \widehat{\theta}_j - t_{n-k-1;\frac{\alpha}{2}} \cdot \sqrt{s_{jj}} \cdot S\right\} = 1 - \alpha.$$

Considering the symmetry of the density of the student distribution we obtain $t_{n-k-1;\frac{\alpha}{2}} = -t_{n-k-1;1-\frac{\alpha}{2}}$, and therefore the equivalent inequalities

$$Pr\left\{\widehat{\theta}_j - t_{n-k-1;1-\frac{\alpha}{2}} \cdot \sqrt{s_{jj}} \cdot S \le \theta_j \le \widehat{\theta}_j + t_{n-k-1;\frac{\alpha}{2}} \cdot \sqrt{s_{jj}} \cdot S\right\} = 1 - \alpha.$$

(2) The confidence interval for σ^2 is obtained using the pivotal quantity $\frac{(n-k-1)S^2}{\sigma^2}$:

$$Pr\left\{\chi^2_{n-k-1;\frac{\alpha}{2}} \le \frac{(n - k - 1)S^2}{\sigma^2} \le \chi^2_{n-k-1;1-\frac{\alpha}{2}}\right\} = 1 - \alpha.$$

Identical transformation of the inequalities yields

$$Pr\left\{\frac{(n-k-1)S^2}{\chi^2_{n-k-1;1-\frac{\alpha}{2}}} \le \sigma^2 \le \frac{(n-k-1)S^2}{\chi^2_{n-k-1;\frac{\alpha}{2}}}\right\} = 1-\alpha.$$

This proves the confidence interval for σ^2.

19.3   General linear models

In generalization of the linear regression model the functional form of

$$\mathbb{E}Y_x = \Psi(x_1, \cdots, x_k; \theta_0, \theta_1, \cdots, \theta_r)$$

for a *general linear model* is assumed to be given by r known functions $f_j(x_1, \cdots, x_k)$, $j = 1(1)r$ in the form of

$$\Psi(x_1, \cdots, x_k; \theta_0, \theta_1, \cdots, \theta_r) = \theta_0 + \sum_{j=1}^{r}\theta_j f_j(x_1, \cdots, x_k) \qquad (19.9)$$

with unknown parameters $\theta_0, \theta_1, \cdots, \theta_r$, and linear independent functions $f_1(\cdot), \cdots, f_r(\cdot)$. The functions $f_j(x_1, \cdots, x_k)$ can be written in compact form as $f_j(\boldsymbol{x})$, and using the notation

$$\boldsymbol{f}(\cdot) = \begin{pmatrix} f_1(\cdot) \\ \vdots \\ f_r(\cdot) \end{pmatrix} \quad \text{and} \quad \mathring{\boldsymbol{f}}(\cdot) = \begin{pmatrix} 1 \\ f_1(\cdot) \\ \vdots \\ f_r(\cdot) \end{pmatrix}$$

$$\boldsymbol{\theta} = (\theta_0, \theta_1, \cdots, \theta_r), \quad \boldsymbol{x} = (x_1, \cdots, x_k)$$

the general linear model (19.9) takes the form

$$\mathbb{E}Y_x = \boldsymbol{\theta}\,\mathring{\boldsymbol{f}}(\boldsymbol{x}).$$

As in the linear regression model usually the variances are assumed to be identical, i.e.

$$\text{Var } Y_x = \sigma^2 \quad \text{for all} \quad \boldsymbol{x}.$$

In order to estimate the parameters $\theta_0, \theta_1, \cdots, \theta_r$ and σ^2 observations are taken, i.e. the data are of the same type as for linear regression models:

$$(x_{i1}, \cdots, x_{ik}; y_i), \quad i = 1(1)n.$$

Based on these data, and using the notation $x_i = (x_{i1}, \cdots, x_{ik})$ the data matrix $\mathcal{F}$ is defined by

$$\mathcal{F} = \begin{pmatrix} 1, & f_1(x_1) & , \cdots, & f_r(x_1) \\ \vdots & \vdots & & \\ 1, & f_1(x_n) & , \cdots, & f_r(x_n) \end{pmatrix}.$$

In order to estimate the regression parameters $\theta_0, \theta_1, \cdots, \theta_r$ again the method of LSS is used, i.e.

$$SS = \sum_{i=1}^{n} \left[y_i - \theta_0 - \sum_{j=1}^{r} \theta_j f_j(x_{i1}, \cdots, x_{ik}) \right] \longrightarrow \min.$$

Similar to linear regression models under the following conditions the estimations $\widehat{\theta}_j$ for the parameters θ_j can be given explicitly. This is the subject of the following theorem:

Theorem 19.4: Let a general linear model (19.9) be given. Furthermore a sample $(x_i, y_i), i = (1)n$, and the following conditions are fulfilled:

(1) $\mathbb{E}Y_x = \theta \, \mathring{f}(x) \quad \forall \quad x$ to be considered.

(2) $\mathrm{Var}\, Y_x = \sigma^2 \, \forall \, x$ to be considered.

(3) $Y_{x_1}, \cdots, Y_{x_n}$ are pairwise uncorrelated.

(4) $\mathrm{rank}\left(\mathcal{F}^T \mathcal{F}\right) = r + 1 < n$.

(5) The space of possible parameters θ is not a hyperplane in $\mathbb{R}^{r+1}$.

Then the following holds:

1) The LSS estimates for the parameters are given by

$$\widehat{\theta}^T = \left(\mathcal{F}^T \mathcal{F}\right)^{-1} \mathcal{F}^T y.$$

2) The estimator $\left(\mathcal{F}^T \mathcal{F}\right)^{-1} \mathcal{F}^T Y$ provides the MVULE for the parameters.

3) The variance–covariance matrix of $\widehat{\theta}$ is given by

$$\mathrm{VCov}\left(\widehat{\theta}\right) = \left(\mathcal{F}^T \mathcal{F}\right)^{-1} \sigma^2.$$

4) An unbiased estimator for σ^2 is given by

$$S^2 = \frac{\left(Y - \mathcal{F}\widehat{\theta}^T\right)^T \left(Y - \mathcal{F}\widehat{\theta}\right)^T}{n - r - 1} = \frac{\sum_{i=1}^{n}\left[Yx_i - \widehat{\theta}_0 - \sum_{j=1}^{r}\widehat{\theta}_j f_j(x_i)\right]}{n - r - 1}.$$

The proof is given in books on regression analysis.

A prediction for the dependent variable y based on the value x for the independent variable is given by

$$\widehat{Y}_x = \Psi(x; \widehat{\theta}) = \widehat{\theta}\, \mathring{f}(x). \tag{19.10}$$

Proposition 19.3: $\widehat{Y}_x$ above is an unbiased estimate for $\mathbb{E}Y_x$, and the variance of this estimate is

$$\operatorname{Var} \widehat{Y}_x = \mathring{f}^T(x)\left(\mathcal{F}^T \mathcal{F}\right)^{-1} \mathring{f}(x)\sigma^2.$$

Proof: $\mathbb{E}\left(\widehat{\theta}\,\mathring{f}(x)\right) = (\mathbb{E}\widehat{\theta})\,\mathring{f}(x) = \theta\,\mathring{f}(x) = \mathbb{E}Y_x$

$$\operatorname{Var} \widehat{Y}_x = \operatorname{Var}\left(\widehat{\theta}\,\mathring{f}(x)\right) = \operatorname{Var}\left(\left[(\mathcal{F}^T\mathcal{F})^{-1}\mathcal{F}^T Y\right]^T \mathring{f}(x)\right)$$

$$= \operatorname{Var}\left(Y^T \underbrace{\mathcal{F}\left[(\mathcal{F}^T\mathcal{F})^{-1}\right]^T \mathring{f}(x)}\right) = \operatorname{Var}\left(Y^T z\right) = \operatorname{Var}\left(\sum_{i=1}^{n} z_i Y_i\right).$$

$$\text{with } z = \begin{pmatrix} z_1 \\ \vdots \\ z_n \end{pmatrix}$$

Since $Y_1, \cdots, Y_n$ are pairwise uncorrelated we obtain $\operatorname{Var}\left(\sum_{i=1}^{n} z_i Y_i\right) =$

$$\sum_{i=1}^{n} z_i^2 \underbrace{\operatorname{Var} Y_i}_{\sigma^2} = z^T z \sigma^2.$$

From $z^T z = \left(\mathcal{F}\left[(\mathcal{F}^T\mathcal{F})^{-1}\right]^T \mathring{f}(x)\right)^T \left(\mathcal{F}\left[(\mathcal{F}^T\mathcal{F})^{-1}\right]^T \mathring{f}(x)\right) =$

$$= \mathring{f}^T(x)\underbrace{\left(\mathcal{F}^T\mathcal{F}\right)^{-1}\mathcal{F}^T\mathcal{F}}_{I_n}\left[(\mathcal{F}^T\mathcal{F})^{-1}\right]^T \mathring{f}(x) = \mathring{f}^T(x)\left[(\mathcal{F}^T\mathcal{F})^{-1}\right]^T \mathring{f}(x) \in \mathbb{R}$$

$$\Rightarrow \left(\mathring{f}^T(x)\left[(\mathcal{F}^T\mathcal{F})^{-1}\right]^T \mathring{f}(x)\right)^T = \mathring{f}^T(x)\left[\mathcal{F}^T\mathcal{F})^{-1}\right]^T \mathring{f}(x)$$

$$\Rightarrow \mathring{f}^T(x)\left[(\mathcal{F}^T\mathcal{F})^{-1}\right]^T \mathring{f}(x) = \mathring{f}^T(x)(\mathcal{F}^T\mathcal{F})^{-1}\mathring{f}(x)$$

$$\Rightarrow \operatorname{Var} \widehat{Y}_x = \mathring{f}^T(x)(\mathcal{F}^T\mathcal{F})^{-1}\mathring{f}(x)\sigma^2.$$

$\square$

19.4 Nonidentical variances

The assumption of identical variances in Theorem 19.1 is sometimes not justified. Moreover the error terms described by stochastic quantities U_i are not always uncorrelated. Therefore the regression model

$$Y = \mathcal{X}\theta + U \quad \text{with} \quad \text{VCov}(U) = I_n \sigma^2$$

is generalized in the following way:

$$\text{VCov}(U) = \sigma^2 \sum$$

where $\sum$ is a given positive definite $(n \times n)$ matrix.

The corresponding linear regression model is

$$\mathbb{E} Y_x = \theta \, \mathring{f}(x),$$

and including the error terms U_x it becomes

$$Y_x = \theta \, \mathring{f}(x) + U_x$$

with $\text{VCov}(U) = \sigma^2 \sum$.

Again the method of LSS is applied to estimate the parameters $\theta_0, \theta_1, \cdots, \theta_r$ from data (x_i, y_i), $i = 1(1)n$. The following theorem is a generalization of the Gauss–Markov theorem, called the Gauss–Markov–Aitken theorem.

Theorem 19.5: Let a general linear model (19.9) be given. Furthermore a sample (x_i, Y_i), $i = 1(1)n$. Assuming the following conditions are fulfilled:

(1) rank $(\mathcal{F}) = r + 1 < n$;

(2) $\text{VCov}(Y) = \sigma^2 \sum$;

(3) the space of possible parameters θ is not a hyperplane in $\mathbb{R}^{r+1}$,

then the following holds:

1) By $\widehat{\theta}^T = \left(\mathcal{F}^T \sum^{-1} \mathcal{F} \right)^{-1} \mathcal{F}^T \sum^{-1} Y$ the unequally determined MVULE for the parameter vector θ are given (called Aitken estimators).

2) The variance–covariance matrix $\text{VCov}(\widehat{\theta})$ of this estimator is given by

$$\text{VCov}(\widehat{\theta}) = \left(\mathcal{F}^T \sum^{-1} \mathcal{F} \right)^{-1} \sigma^2.$$

3) An unbiased estimator for the parameter σ^2 is given by

$$S^2 := \frac{\left(\boldsymbol{Y} - \mathcal{F}\widehat{\boldsymbol{\theta}}^T\right)^T \Sigma^{-1} \left(\boldsymbol{Y} - \mathcal{F}\widehat{\boldsymbol{\theta}}^T\right)}{n - r - 1}.$$

For the proof see books on regression analysis, for example Schönfeld (1969).

19.5   Problems

(a) Explain that polynomial regression functions are special forms of linear regression.

(b) What kind of functions $f_j(\cdot)$ in the general linear model generate the classical linear regression model?

(c) Assuming the random variables $Y_{\boldsymbol{x}_i}$ are all normally distributed. What does assumption (3) in Theorem 19.1 imply?

20

Regression models and fuzzy data

As mentioned at the beginning of this book, all measurement results of continuous quantities are not precise numbers. Therefore regression models also have to take care of fuzzy data. A sample therefore is given by n vectors of fuzzy numbers, i.e.

$$\left(x_{i1}^*, \cdots, x_{ik}^*; y_i^*\right), \ i = 1(1)n.$$

Another kind of fuzzy data are given if the vector of independent variables is a fuzzy vector x_i^*. Then the sample takes the following form:

$$\left(x_i^*; y_i^*\right), \ i = 1(1)n.$$

Remark 20.1: It is important to note that the fuzzy vector x_i^* is essentially different from the vector of fuzzy numbers $\left(x_{i1}^*, \cdots, x_{ik}^*\right)$. The former is defined by a vector-characterizing function, whereas the latter is defined by a vector of characterizing functions.

In applications both kinds of fuzzy data appear.

Another kind of fuzziness in this context is the fuzziness of parameters in regression models.

There are several situations possible if fuzziness is taken into account in regression models:

(a) The parameters and the independent variables x_i are assumed to be classical real valued, but the dependent variable is fuzzy, i.e. y_i^* are fuzzy numbers.

(b) The independent variables x_i as well as the values of the dependent variables y_i are classical real numbers but the parameters are fuzzy numbers θ_j^*, i.e.

$$\mathbb{E}Y_x = \theta_0^* \oplus \theta_1^* x_1 \oplus \cdots \oplus \theta_k^* x_k.$$

Here $\oplus$ denotes the generalized addition operation for fuzzy numbers.

(c) The values of the input variables x_i are fuzzy numbers x_i^* and all other quantities are classical real numbers.

(d) The input variables as well as the output variables are fuzzy, and the parameters are classical real valued, i.e.

$$\mathbb{E}Y_x = \theta_0 \oplus \theta_1 x_1^* \oplus \cdots \oplus \theta_k x_k^*,$$

and the data set is $\left(x_{i1}^*, \cdots, x_{ik}^*; y_i^*\right)$.

(e) All considered quantities are fuzzy, i.e.

$$\mathbb{E}Y_x = \theta_0^* \oplus \theta_1^* \odot x_1^* \oplus \cdots \oplus \theta_k^* \odot x_k^*.$$

The above described possibilities apply also to general linear models. All the above situations can be adapted.

There are different approaches for the generalization of estimation procedures in the case of fuzzy data. The most direct approach is based on the extension principle. This is described in the next section.

20.1   Generalized estimators for linear regression models based on the extension principle

Depending on the kind of available data generalized estimators for the parameters can be obtained in the following way:

ad (a): $\mathbb{E}Y_x = \theta_0 + \theta_1 x_1 + \cdots + \theta_k x_k$ fuzzy dependent variables y_i^*:

For data $(x_{1i}, \cdots, x_{ki}; y_i^*)$, $i = 1(1)n$ the fuzzy values y_i^* are characterized by their characterizing functions $\eta_i(\cdot)$, $i = 1(1)n$.

In this situation the estimators $\widehat{\theta}_j$ for θ_j, $j = 1(1)k$ can be generalized in the following way:

In order to apply the extension principle first the fuzzy data $y_1^*, \cdots, y_n^*$ have to be combined to form a fuzzy vector $\boldsymbol{y}^*$ in $\mathbb{R}^n$.

The vector-characterizing function $\zeta(\cdot, \cdots, \cdot)$ of the combined fuzzy vector $\boldsymbol{y}^*$ is obtained from the characterizing functions $\eta_1(\cdot), \cdots, \eta_n(\cdot)$ by application of the minimum t-norm, i.e. the values $\zeta(y_1 \cdots, y_n)$ are defined by

$$\zeta(y_1, \cdots, y_n) := \min\{\eta_1(y_1), \cdots, \eta_n(y_n)\} \quad \forall \quad (y_1, \cdots, y_n) \in \mathbb{R}^n.$$

The estimators $\widehat{\theta}_j$ of the regression parameters θ_j can now be generalized.

Using the notation $\boldsymbol{x}_i = (x_{i1}, \cdots, x_{ik})$ for $i = 1(1)n$ the estimators $\widehat{\theta}_j$ for θ_j from Theorem 19.1 can be written as

$$\widehat{\theta}_j = \widehat{\theta}_j(\boldsymbol{x}_1, \cdots, \boldsymbol{x}_n; y_1, \cdots, y_n).$$

Now the membership functions $\vartheta_j(\cdot)$ of the generalized (fuzzy) estimator $\widehat{\theta}_j^*$ based on the fuzzy data $y_1^*, \cdots, y_n^*$, i.e.

$$\widehat{\theta}_j^* = \widehat{\theta}_j\left(\boldsymbol{x}_1, \cdots, \boldsymbol{x}_n; y_1^*, \cdots, y_n^*\right)$$

are given by their values $\vartheta_j(z)$ in the following way:

$$\vartheta_j(z) = \left\{ \begin{array}{l} \sup\left\{\zeta(y_1, \cdots, y_n): \widehat{\theta}_j(\boldsymbol{x}_1, \cdots, \boldsymbol{x}_n; y_1, \cdots, y_n) = z\right\} \\ \quad \text{if } \exists (y_1, \cdots, y_n): \widehat{\theta}_j(\boldsymbol{x}_1, \cdots, \boldsymbol{x}_n; y_1, \cdots, y_n) = z \\[2mm] 0 \quad \text{if } \not\exists(y_1, \cdots, y_n): \widehat{\theta}_j(\boldsymbol{x}_1, \cdots, \boldsymbol{x}_n; y_1, \cdots, y_n) = z \end{array} \right\} \forall z \in \mathbb{R}.$$

Remark 20.2: By the continuity of the functions $\widehat{\theta}_j(\boldsymbol{x}_1, \cdots, \boldsymbol{x}_n; y_1, \cdots, y_n)$ the membership functions $\vartheta_j(\cdot)$ are characterizing functions of fuzzy intervals if all functions $\eta_j(\cdot)$ are characterizing functions of fuzzy intervals y_i^*.

A generalized (fuzzy) estimator of the variance σ^2 is given by the fuzzy element σ^{2*} of $\mathbb{R}$ whose membership function $\psi(\cdot)$ is given by its values

$$\psi(z) = \left\{ \begin{array}{l} \sup\left\{\zeta(y_1, \cdots, y_n): S^2(\boldsymbol{x}_1, \cdots, \boldsymbol{x}_n; y_1, \cdots, y_n) = z\right\} \\ \quad \text{if } \exists (y_1, \cdots, y_n): S^2(\boldsymbol{x}_1, \cdots, \boldsymbol{x}_n; y_1, \cdots, y_n) = z \\[2mm] 0 \quad \text{if } \not\exists (y_1, \cdots, y_n): S^2(\boldsymbol{x}_1, \cdots, \boldsymbol{x}_n; y_1, \cdots, y_n) = z \end{array} \right\} \forall z \in \mathbb{R}.$$

where S^2 is the function from conclusion 4) in Theorem 19.1

Remark 20.3: The generalized regression function for fuzzy data is a fuzzy valued function of the classical variables $x_1, \cdots, x_n$. This fuzzy valued function

$$y^* = \widehat{\theta}_0^* \oplus \widehat{\theta}_1^* x_1 \oplus \cdots \oplus \widehat{\theta}_k^* x_k$$

can be characterized by the lower and upper δ-level functions $\underline{y}_\delta(x_1, \cdots, x_k)$ and $\overline{y}_\delta(x_1, \cdots, x_k)$, respectively.

An example for $k = 1$ is a generalized regression line

$$\mathbb{E}Y_x = \theta_0 + \theta_1 x \qquad \text{for} \quad x \in \mathbb{R}.$$

For fuzzy data given in Table 20.1 the generalized estimators $\hat{a}^*$ and $\hat{b}^*$ for the regression parameters a and b in

$$\mathbb{E}Y_x = a + bx$$

Table 20.1   Fuzzy data.

x_i	$y_i^* = < m_i,\ l_i,\ r_i >_L$
6.15	$< 2.5,\ 0.17,\ 0.40 >_L$
9.00	$< 2.0,\ 0.17,\ 0.45 >_L$
12.00	$< 3.0,\ 0.30,\ 0.50 >_L$
12.15	$< 2.5,\ 0.25,\ 0.53 >_L$
15.00	$< 3.0,\ 0.13,\ 0.46 >_L$
15.19	$< 3.5,\ 0.35,\ 0.63 >_L$
17.55	$< 3.5,\ 0.27,\ 0,72 >_L$
20.48	$< 4.0,\ 0.32,\ 0.61 >_L$

are given by

$$\hat{a}^* = < 1.3079,\ 0.9208,\ 1.0510 >_L$$
$$\hat{b}^* = < 0.1259,\ 0.0707,\ 0.0820 >_L .$$

For the generalized regression function

$$\mathbb{E}Y_x = \hat{a}^* \oplus \hat{b}^* x$$

we obtain the fuzzy line $\widehat{f}^*(x)$ with

$$\widehat{f}^*(x) = < 1.3079,\ 0.9208,\ 1.0510 >_L \oplus < 0.1259,\ 0.0707,\ 0.0820 >_L x.$$

In Figure 20.1 the fuzzy data $y_1^*, \cdots, y_8^*$ as well as the fuzzy regression line are depicted. The solid line is the ($\delta = 1$) level curve of $\widehat{f}^*(\cdot)$. The dashed lines are the connections of the lower and upper ends of the supports of $\widehat{f}^*(\cdot)$.

ad (c): If the values of input variables x_i are fuzzy numbers, again the generalized estimators for the regression parameters are becoming fuzzy.

Let $\xi_i(\cdot), i = 1(1)n$ denote the vector-characterizing functions of the fuzzy vectors $x_i^* = (x_{i1}, \cdots, x_{ik})^*$.

In order to generalize the estimators $\widehat{\theta}_j$ to the situation of fuzzy data, the fuzzy vectors x_i^* have to be combined into a fuzzy element of the Euclidean space $\mathbb{R}^{nk}$. This fuzzy element x^* of $\mathbb{R}^{nk}$ has a vector-characterizing function $\zeta : \mathbb{R}^{nk} \to [0; 1]$ whose values $\zeta(x)$ are defined by the minimum t-norm:

Obeying $x = (x_1, \cdots, x_n)$ the values $\zeta(x)$ are given by

$$\zeta(x_1, \cdots, x_n) := \min\{\xi_1(x_1), \cdots, \xi_n(x_n)\} \quad \forall \quad x_i \in \mathbb{R}^k.$$

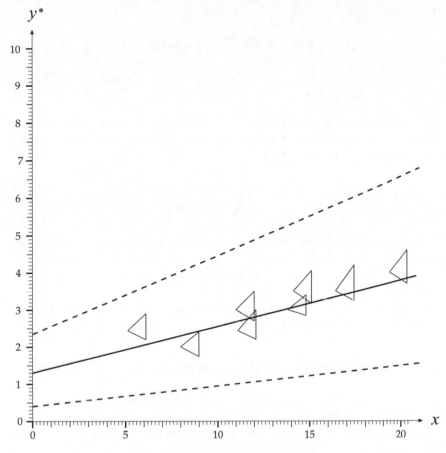

Figure 20.1   Linear regression line.

Based on this the generalized estimators $\widehat{\theta}_j^{\,*}$ are obtained by the extension principle. The membership function $\phi_j(\cdot)$ of $\widehat{\theta}_j^{\,*}$ is given for $j = 1(1)k$ by

$$
\phi_j(z) := \left\{
\begin{array}{ll}
\sup \left\{ \zeta(\boldsymbol{x}) : \widehat{\theta}_j(\boldsymbol{x}; y_1, \cdots, y_n) = z \right\} \\
\quad \text{if } \exists\, \boldsymbol{x} \in \mathbb{R}^{nk} : \widehat{\theta}_j(\boldsymbol{x}; y_1, \cdots, y_n) = z \\[2ex]
0 \quad \text{if } \nexists\, \boldsymbol{x} \in \mathbb{R}^{nk} : \widehat{\theta}_j(\boldsymbol{x}; y_1, \cdots, y_n) = z
\end{array}
\right\} \qquad \forall\, z \in \mathbb{R}.
$$

ad (d): If the input variables as well as the output variables are fuzzy, the fuzziness of all of these variables has to be combined in order to apply the extension principle. This is done by using the minimum t-norm.

Applying the notation of case (c) above, and denoting the characterizing function of y_i^* by $\eta_i(\cdot)$ for $i = 1(1)n$, the combined fuzziness is contained in the fuzzy element $\boldsymbol{t}^*$ of $\mathbb{R}^{n(k+1)}$ whose vector-characterizing function $\tau(\cdot, \cdots, \cdot)$ is given by its values

$\tau(t)$, defined by

$$\tau(x_1, \cdots, x_n, y_1, \cdots, y_n) := \min \{\xi_1(x_1), \cdots, \xi_n(x_n), \eta_1(y_1), \cdots, \eta_n(y_n)\}$$

$$\forall \begin{cases} x_i \in \mathbb{R}^k \\ y_i \in \mathbb{R} . \end{cases}$$

Based on this fuzzy element t^* in $\mathbb{R}^{n(k+1)}$, which is a fuzzy vector, the estimators $\widehat{\theta}_j$ for the regression parameters can be generalized. The characterizing function $\phi_j(\cdot)$ of the fuzzy estimator $\hat{\theta}_j^*$ is given by

$$\phi_j(z) := \begin{cases} \sup \left\{\tau(t) : \widehat{\theta}_j(t) = z\right\} & \text{if} \quad \exists\, t \in \mathbb{R}^{n(k+1)} : \widehat{\theta}_j(t) = z \\ 0 & \text{if} \quad \nexists\, t \in \mathbb{R}^{n(k+1)} : \widehat{\theta}_j(t) = z \end{cases} \forall z \in \mathbb{R}.$$

The resulting estimates $\hat{\theta}_j^*$ are fuzzy elements of $\mathbb{R}$. In the case of continuous functions $\widehat{\theta}_j(\cdot)$ the fuzzy estimators are fuzzy numbers.

ad (e): In this case the estimation for the parameters is done in the same way as before in case (d).

For the model (b) classical estimation procedures would be suitable. The fuzzy parameters θ_j^* can be obtained by fuzzification of the standard estimates $\widehat{\theta}_j$.

20.2   Generalized confidence regions for parameters

For standard regression models with existing confidence function $\kappa_j(Y_1, \cdots, Y_n)$ for parameters θ_j in the case of fuzzy data $y_1^*, \cdots, y_n^*$ corresponding generalized (fuzzy) confidence regions $\Theta_{1-\alpha}^*$ can be obtained by application of the method from Section 12.2.

20.3   Prediction in fuzzy regression models

Depending on the assumed regression model, predictions of dependent values for values of the independent variable where no data are given can be obtained using the generalized algebraic operations for fuzzy quantities.

The predictions depend on the different cases (a)–(e) given at the start of the chapter.

ad (a): In this case the estimated parameters are fuzzy numbers $\hat{\theta}_j^*$ and the predicted value $y^*(x) = \hat{\theta}_0^* \oplus \hat{\theta}_1^* x_1 \oplus \cdots \oplus \hat{\theta}_k^* x_k$. Therefore $y^*(x)$ is a fuzzy number whose characterizing function is obtained by the generalized arithmetic operations.

ad (b): For the prediction the situation is essentially the same as in (a).

ad (c): By the estimation procedure for θ_j the estimators are fuzzy values $\hat{\theta}_j^*$, and the prediction equation reads

$$y^*(x^*) = \hat{\theta}_0^* \oplus \hat{\theta}_1^* \odot x_1^* \oplus \cdots \oplus \hat{\theta}_k^* \odot x_k^*.$$

Therefore fuzzy multiplication as well as fuzzy addition has to be applied here.

ad (d): Here the prediction equation based on observed data is the same as in case (c).

ad (e): Again the prediction equation has the form from case (c).

Example 20.1 For applications triangular fuzzy numbers are not sufficient. Frequently trapezoidal fuzzy numbers are more realistic. In this situation for the regression model $y^*(x^*) = \hat{\theta}_0^* \oplus \hat{\theta}_1^* \odot x_1^* \oplus \cdots \oplus \hat{\theta}_k^* \odot x_k^*$ the resulting fuzzy values $y^*(x^*)$ are not necessarily trapezoidal fuzzy numbers but fuzzy intervals.

20.4   Problems

(a) For regression line $y(x) = \theta_0 + \theta_1 x$ and two fuzzy observations (x_1^*, y_1^*) and (x_2^*, y_2^*) explain how to calculate the fuzzy estimators $\hat{\theta}_0^*$ and $\hat{\theta}_1^*$.

(b) Calculate the fuzzy prediction $y^*(x^*) = \hat{\theta}_0^* \oplus \hat{\theta}_1^* \odot x^*$ for the fuzzy value x^* of the independent variable which is a fuzzy number.

21

Bayesian regression analysis

In Bayesian inference the parameters θ are also considered as stochastic quantities $\tilde{\theta}$ with corresponding probability distribution $\pi(\cdot)$, called *a priori distribution*. In the case of continuous parameters the a priori distribution is determined by a probability density $\pi(\cdot)$ on the parameter space

$$\Theta = \{\theta : \theta \text{ possible parameter value}\}.$$

Then $\pi(\cdot)$ is a probability density on the parameter space, called *a priori density*.

The stochastic model for the dependent variable y is $Y_x \sim f_x(\cdot \mid \theta)$.

21.1 Calculation of a posteriori distributions

For observed data (x_i, y_i), $i = 1(1)n$ the likelihood function $\ell(\theta; (x_1, y_1), \cdots, (x_n, y_n))$ is given by

$$\ell(\theta; (x_1, y_1), \cdots, (x_n, y_n)) = \prod_{i=1}^{n} f_{x_i}(y_i \mid \theta)$$

in the case of independent observations $y_1, \cdots, y_n$.

The a posteriori density $\pi(\cdot \mid (x_1, y_1), \cdots, (x_n, y_n))$ of $\tilde{\theta}$ is given by Bayes' theorem which reads here

$$\pi(\theta \mid (x_1, y_1), \cdots, (x_n, y_n)) \propto \pi(\theta) \cdot \ell(\theta; (x_1, y_1), \cdots, (x_n, y_n)) \quad \forall \theta \in \Theta.$$

In extended form, Bayes' theorem takes the following form:

$$\pi\left(\boldsymbol{\theta} \mid (\boldsymbol{x}_1, y_1), \cdots, (\boldsymbol{x}_n, y_n)\right) = \frac{\pi(\boldsymbol{\theta}) \cdot \ell\left(\boldsymbol{\theta}; (\boldsymbol{x}_1, y_1), \cdots, (\boldsymbol{x}_n, y_n)\right)}{\int\limits_{\Theta} \pi(\boldsymbol{\theta}) \cdot \ell\left(\boldsymbol{\theta}; (\boldsymbol{x}_1, y_1), \cdots, (\boldsymbol{x}_n, y_n)\right) d\boldsymbol{\theta}} \quad \forall \; \boldsymbol{\theta} \in \Theta.$$

Here the parameter $\boldsymbol{\theta} = (\theta_1, \cdots, \theta_k) \in \mathbb{R}^k$ can be k-dimensional. Then the above integral is

$$\int\limits_{\Theta} \pi(\theta_1, \cdots, \theta_k) \cdot \ell\left(\theta_1, \cdots, \theta_k; (\boldsymbol{x}_1, y_1), \cdots, (\boldsymbol{x}_n, y_n)\right) d\theta_1 \cdots d\theta_k.$$

The a posteriori density carries all information concerning the parameter vector $\boldsymbol{\theta} = (\theta_1, \cdots, \theta_k)$ which comes from the a priori distribution and the data.

21.2   Bayesian confidence regions

The a posteriori distribution can be used to construct Bayesian confidence regions, called *highest a posteriori density regions*, abbreviated as HPD-regions for the parameter vector.

21.3   Probabilities of Hypotheses

Another way of using a posteriori distributions is for the calculation of a posteriori probabilities of statistical hypotheses

$$\mathcal{H}_0 : \boldsymbol{\theta} \in \Theta_0 \subset \Theta.$$

The a posteriori probability of the hypothesis $\mathcal{H}_0$ is the probability of Θ_0 based on the a posteriori density $\pi\left(\cdot \mid (\boldsymbol{x}_1, y_1), \cdots, (\boldsymbol{x}_n, y_n)\right)$, i.e.

$$Pr\left(\Theta_0 \mid (\boldsymbol{x}_1, y_1), \cdots, (\boldsymbol{x}_n, y_n)\right) = \int\limits_{\Theta_0} \pi\left(\boldsymbol{\theta} \mid (\boldsymbol{x}_1, y_1), \cdots, (\boldsymbol{x}_n, y_n)\right) d\boldsymbol{\theta}.$$

Depending on the value of $Pr\left(\Theta_0 \mid (\boldsymbol{x}_1, y_1), \cdots, (\boldsymbol{x}_n, y_n)\right)$ the hypothesis $\mathcal{H}_0$ is accepted or rejected.

21.4   Predictive distributions

An elegant way of using the a posteriori distribution $\pi\left(\cdot \mid (\boldsymbol{x}_1, y_1), \cdots, (\boldsymbol{x}_n, y_n)\right)$ is the predictive distribution of Y_x based on the observed data $(\boldsymbol{x}_1, y_1), \cdots, (\boldsymbol{x}_n, y_n)$. This *predictive distribution* is calculated as the marginal distribution of Y_x of the stochastic vector $\left(Y_x, \widetilde{\boldsymbol{\theta}}\right)$.

For continuous variables Y_x and continuous parameter θ with probability density $f_x(\cdots \mid \theta)$ of Y_x and a posteriori density $\pi(\cdot \mid (x_1, y_1), \cdots, (x_n, y_n))$ of $\widetilde{\theta}$ the joint density $g(\cdot, \cdot)$ of $(Y_x, \widetilde{\theta})$ is given by its values

$$g(y, \theta) = f_x(y \mid \theta) \cdot \pi(\theta \mid (x_1, y_1), \cdots, (x_n, y_n)).$$

Therefore the predictive distribution of Y_x is given by its probability density $p_x(\cdot \mid (x_1, y_1), \cdots, (x_n, y_n))$ which is the marginal density of Y_x, given by its values

$$p_x(y \mid (x_1, y_1), \cdots, (x_n, y_n)) = \int_{\Theta} f_x(y \mid \theta) \cdot \pi(\theta \mid (x_1, y_1), \cdots, (x_n, y_n)) d\theta.$$

The density $p_x(\cdot \mid (x_1, y_1), \cdots, (x_n, y_n))$ is denoted as the *predictive density* of Y_x.

The predictive distribution of Y_x can be used to calculate Bayesian *predictive intervals* for Y_x.

Based on the predictive density a predictive interval $[a; b]$ for Y_x with probability of coverage $1 - \alpha$ is defined by

$$\int_a^b p_x(y \mid (x_1, y_1), \cdots, (x_n, y_n)) dy = \Pr(Y_x \in [a; b]) = 1 - \alpha.$$

Special predictive intervals are HPD intervals for Y_x which make the predictive density maximal in the interval $[a; b]$.

21.5 A posteriori Bayes estimators for regression parameters

In case one wants to have point estimates for the regression parameters $\theta_1, \cdots, \theta_k$ this is possible by using the marginal distributions of the individual parameters. Let $\pi_j(\cdot)$ be the marginal distribution of $\widetilde{\theta}_j$ calculated from the a posteriori density $\pi(\cdot \mid D)$, then the a posteriori Bayes estimator for θ_j is the expectation of $\widetilde{\theta}_j$, i.e.

$$\widehat{\theta}_j := \int_{\Theta_j} \theta \cdot \pi_j(\theta) d\theta.$$

The a posteriori Bayes estimator for the parameter vector $\theta = (\theta_1, \cdots, \theta_k)$ is given by

$$\widehat{\theta} = (\widehat{\theta}_1, \cdots, \widehat{\theta}_k).$$

Remark 21.1: Another possibility for point estimates for the θ_j would be the median of the marginal distribution of $\widetilde{\theta}_j$.

21.6   Bayesian regression with Gaussian distributions

For linear regression models with vector independent variable $x = (x_1, \cdots, x_k)$,

$$\mathbb{E}Y_x = \theta_0 + \sum_{j=1}^{k} \theta_j x_j$$

and homoscedastic Gaussian distributions of the dependent variable Y_x the likelihood function takes the following form:

If the variance σ^2 is known we obtain

$$\ell\left(\theta_0, \theta_1, \cdots, \theta_k; (x_1, y_1), \cdots, (x_n, y_n)\right) = \prod_{i=1}^{n} f_{x_i}(y_i \mid \theta_0, \cdots, \theta_k)$$

$$= \prod_{i=1}^{n} \frac{1}{\sqrt{2\pi\sigma^2}} e^{-\frac{1}{2}\frac{\left(y_i - \theta_0 - \sum_{j=1}^{k}\theta_j x_j\right)^2}{\sigma^2}}.$$

For unknown variance σ^2 the likelihood is a function of σ^2 also,

$$\ell\left(\theta_0, \cdots, \theta_k, \sigma^2; (x_1, y_1), \cdots, (x_n, y_n)\right) = \prod_{i=1}^{n} \frac{1}{\sqrt{2\pi\sigma^2}} e^{-\frac{\left(y_i - \theta_0 - \sum_{j=1}^{k}\theta_j x_j\right)^2}{2\sigma^2}},$$

and the a priori density has to be a function of $\theta_0, \cdots, \theta_k$ and σ^2, i.e. $\pi(\theta_0, \theta_1, \cdots, \theta_k, \sigma^2)$.

The a posteriori density is given by Bayes' theorem:

$$\pi\left(\theta_0, \cdots \theta_k, \sigma^2 \mid (x_1, y_1), \cdots, (x_n, y_n)\right)$$

$$= \frac{\pi(\theta_0, \cdots, \theta_k, \sigma^2) \cdot \ell\left(\theta_0, \cdots, \theta_k, \sigma^2; (x_1, y_1), \cdots, (x_n, y_n)\right)}{\int\limits_{\Theta \times (0,\infty)} \pi(\theta_0, \cdots, \theta_k, \sigma^2) \cdot \ell\left(\theta_0, \cdots, \theta_k, \sigma^2; (x_1, y_1), \cdots, (x_n, y_n)\right) d\theta_0 \cdots d\theta_k d\sigma^2}.$$

21.7   Problems

(a) Assume the probability density $f_x(\cdot \mid \boldsymbol{\theta})$ is determined by a parameter vector $\boldsymbol{\psi}$ which is depending on the regression parameters $\boldsymbol{\theta} = (\theta_0, \theta_1, \cdots, \theta_k)$, i.e. $f_{x_i}(\cdot \mid \boldsymbol{\theta}) = g(\cdot \mid \boldsymbol{\psi}(\boldsymbol{\theta}, x_i))$.

Which form has the likelihood function for observed data $(x_1, y_1), \cdots, (x_n, y_n)$? How does Bayes' theorem look in this situation?

(b) Let the dependent quantity Y_x have an exponential distribution Ex_τ with density $f(y \mid \tau) = \frac{1}{\tau} e^{-1/\tau} I_{(0,\infty)}(y)$ with $\tau > 0$. Let $\boldsymbol{\theta} = (\theta_0, \theta_1)^T$. Assuming a regression line $\theta_0 + \theta_1 \cdot x$ for the mean value τ, i.e. $\tau(x) = \theta_0 + \theta_1 \cdot x$ and there-fore $f_x(y \mid \boldsymbol{\theta}) = \frac{1}{\theta_0 + \theta_1 x} \exp \left\{ -\frac{y}{\theta_0 + \theta_1 x} \right\} I_{(0,\infty)}(y)$, calculate the likelihood function and write down Bayes' theorem for this special case for data points $(x_1, y_1), \cdots, (x_n, y_n)$ in the plane $\mathbb{R}^2$.

22

Bayesian regression analysis and fuzzy information

In Bayesian regression analysis two kinds of fuzziness are present: Fuzziness of a priori distributions and fuzziness of data. In this chapter we assume that fuzzy a priori distributions are given by fuzzy densities $\pi^*(\cdot)$ with corresponding δ-level functions $\underline{\pi}_\delta(\cdot)$ and $\overline{\pi}_\delta(\cdot)$, $\delta \in (0; 1)$. The data are given in the form of n vectors of fuzzy numbers, i.e.

$$\left(x_{i\,1}^*, \cdots, x_{i\,k}^*; y_i^*\right), \quad i = 1(1)n,$$

or in the form of fuzzy vectors for the independent variable $x = (x_1, \cdots, x_k)$, and fuzzy number y_i^* for the dependent variable y:

$$\left(x_i^*; y_i^*\right), \quad i = 1(1)n$$

where x_i^* is a k-dimensional fuzzy vector.

In order to apply Bayes' theorem first the likelihood function has to be generalized for fuzzy data. In accordance with Chapter 15 the combined fuzzy sample has to be formed.

In the first situation of data $\left(x_{i\,1}^*, \cdots, x_{i\,k}^*, y_i^*\right)$, $i = 1(1)n$ the combined fuzzy sample z^* is an $n \times (k + 1)$-dimensional fuzzy vector whose vector-characterizing function $\zeta\,(x_{11}, \cdots, x_{nk}, y_1, \cdots, y_n)$ is formed by the minimum t-norm, i.e.

$$\zeta\,(x_{11}, \cdots, x_{nk}, y_1, \cdots, y_n) = \min\left\{\xi_{ij}(x_{ij}), \eta_i(y_i): \ i = 1(1)n, j = 1(1)k\right\}$$

where $\xi_{ij}(\cdot)$ are the characterizing functions of the fuzzy values x_{ij}^*, and $\eta_i(\cdot)$ are the characterizing functions of the fuzzy values y_i^* of the dependent variable.

Statistical Methods for Fuzzy Data   Reinhard Viertl
© 2011 John Wiley & Sons, Ltd

Based on the vector-characterizing function $\zeta(\cdot, \cdots, \cdot)$ of the combined fuzzy sample z^* the fuzzy value of the generalized likelihood function $\ell^*\left(\boldsymbol{\theta}; x_{11}^*, \cdots, x_{nk}^*, y_1^*, \cdots, y_n^*\right)$ is obtained by the extension principle. In order to apply the generalized Bayes' theorem the δ-level functions $\underline{\ell}_\delta(\boldsymbol{\theta}; z^*)$ and $\overline{\ell}_\delta(\boldsymbol{\theta}; z^*)$ of the fuzzy likelihood $\ell^*(\boldsymbol{\theta}; z^*)$ are used. Here $\boldsymbol{\theta} = (\theta_0, \theta_1, \cdots, \theta_k)$, and the fuzzy a priori density $\pi^*(\boldsymbol{\theta})$ has also to be given by its δ-level functions $\underline{\pi}_\delta(\cdot)$ and $\overline{\pi}_\delta(\cdot)$, $\underline{\ell}_\delta(\cdot; z^*)$ and $\overline{\ell}_\delta(\cdot; z^*), \forall \delta \in (0; 1]$. Now the generalized Bayes' theorem can be applied in order to obtain the fuzzy a posteriori density $\pi^*(\cdot \mid z^*)$ via its δ-level functions $\underline{\pi}_\delta(\cdot \mid z^*)$ and $\overline{\pi}_\delta(\cdot \mid z^*) \ \forall \delta \in (0, 1]$.

This fuzzy a posteriori density carries the information from the a priori distribution and the data.

22.1 Fuzzy estimators of regression parameters

Based on the fuzzy a posteriori density $\pi^*(\cdot \mid z^*)$ on the parameter space $\Theta \subseteq \mathbb{R}^r$ generalized point estimators for the parameter $\boldsymbol{\theta} = (\theta_1, \cdots, \theta_r)$ can be calculated.

Without any loss function the most popular method is to calculate the generalized expectation of $\widetilde{\theta}_j$ with respect to $\pi^*(\cdot \mid z^*)$; i.e.

$$\widehat{\theta}_j^* = \mathbb{E}_{\pi^*(\cdot|z^*)}\widetilde{\theta}_j = \int_\Theta \theta_j \cdot \pi_j^*(\boldsymbol{\theta} \mid z^*)d\boldsymbol{\theta} = \int_\mathbb{R} \theta_j \cdot \pi_j^*(\theta_j \mid z^*)d\theta_j$$

where $\pi_j^*(\cdot \mid z^*)$ is the fuzzy marginal density

$$\pi_j^*(\theta_j \mid z^*) = \int_{\mathbb{R}^{r-1}} \pi^*(\theta_1, \cdots, \theta_r \mid z^*) \, d\theta_1 \cdots d\theta_{j-1} \, d\theta_{j+1} \cdots d\theta_r.$$

The characterizing function $\psi_j(\cdot)$ of $\widehat{\theta}_j^*$ is obtained via the construction lemma for characterizing functions. The nested family of subsets of $\mathbb{R}$, $\left(A_\delta; \delta \in (0; 1]\right)$ is given by

$$A_\delta := \left[\underline{A}_\delta; \overline{A}_\delta\right] \quad \forall \delta \in (0; 1],$$

where

$$\overline{A}_\delta := \int_\mathbb{R} \theta_j \cdot \overline{\pi}_{j,\delta} \left(\theta_j \mid z^*\right) d\theta_j$$

and

$$A_\delta := \int_{\mathbb{R}} \theta_j \cdot \underline{\pi}_{j,\delta} \left(\theta_j \mid z^* \right) d\theta_j.$$

Then $\psi_j(\cdot)$ of $\hat{\theta}_j^*$ is given by its values

$$\psi_j(x) = \sup \left\{ \delta \cdot I_{[\underline{A}_\delta;\overline{A}_\delta]}(x) \colon \ \delta \in [0;1] \right\} \quad \forall \, x \in \mathbb{R}.$$

Another possibility is to calculate the generalized (fuzzy) median of the fuzzy marginal distribution $\pi_j^*(\cdot \mid z^*)$. The fuzzy median x_{med}^* is obtained from the fuzzy cumulative distribution function $F_j^*(\cdot)$ coresponding to $\pi_j^*(\cdot \mid z^*)$. The fuzzy valued function $F_j^*(\cdot)$ is defined via its δ-cuts $C_\delta[F_j^*(\cdot)]$ which are defined by the δ-level functions of $\pi_j^*(\cdot \mid z^*)$, i.e. $\underline{\pi}_{j,\delta}(\cdot)$ and $\overline{\pi}_{j,\delta}(\cdot)$ for all $\delta \in (0;1]$,

$$\overline{F}_{j,\delta}(x) := \sup \left\{ \int_{-\infty}^{x} f(t)\, dt \colon f(\cdot) \in \mathcal{S}_\delta \right\}$$

$$\underline{F}_{j,\delta}(x) := \inf \left\{ \int_{-\infty}^{x} f(t)\, dt \colon f(\cdot) \in \mathcal{S}_\delta \right\}$$

where

$$\mathcal{S}_\delta := \left\{ f(\cdot) \colon f(\cdot) \text{ density obeying } \underline{\pi}_{j,\delta}(x) \le f(x) \le \overline{\pi}_{j,\delta}(x) \, \forall \, x \in \mathbb{R} \right\}.$$

The generalized median x_{med}^* of $F_j^*(\cdot)$ is the fuzzy number whose characterizing function $\varphi(\cdot)$ is constructed from the nested family of intervals $\left(B_\delta; \delta \in (0;1] \right)$, given by $B_\delta = \left[\underline{B}_\delta; \overline{B}_\delta \right]$ defined by

$$\left. \begin{aligned} \overline{B}_\delta &:= \underline{F}_{j,\delta}^{-1}(0.5) \\[2mm] \underline{B}_\delta &:= \overline{F}_{j,\delta}^{-1}(0.5) \end{aligned} \right\} \quad \forall \, \delta \in (0;1] :$$

$$\varphi(x) := \sup \left\{ \delta \cdot I_{[\underline{B}_\delta;\overline{B}_\delta]}(x) \colon \ \delta \in [0;1] \right\} \quad \forall \, x \in \mathbb{R}.$$

For loss functions $L(\cdot, \cdot)$ describing the loss which is obtained when the parameter value is θ and this is estimated by $\widehat{\theta}$, i.e. $L(\theta, \widehat{\theta})$, the corresponding estimator is the value θ_B for which the expectation

$$\mathbb{E}_\pi L(\widetilde{\theta}, \theta) \quad \text{is minimized.}$$

In the case of fuzzy loss functions and fuzzy a posteriori distribution $\pi^*(\cdot \mid z^*)$ a fuzzy expectation is obtained. For details see Sections 18.2, 18.3 and 18.4.

22.2   Generalized Bayesian confidence regions

The generalized (fuzzy) estimators $\widehat{\theta}_j^*$ for the regression parameters θ_j are already fuzzy subsets of the parameter space, generalizing point estimators from standard statistics.

In order to obtain generalized confidence sets for regression parameters, based on the explanations in Chapters 16 and 21 fuzzy confidence sets for the regression parameters can be obtained.

22.3   Fuzzy predictive distributions

Predictive distributions for the dependent variable y in a Bayesian regression model are described in Section 21.4.

In the case of fuzzy data the a posteriori distribution is a fuzzy density $\pi^*(\theta \mid z^*)$.

The predictive density $p_x(\cdot \mid z)$ from Section 21.4 has to be generalized to the situation of fuzzy data z^*. Therefore the definition of the predictive density from Section 21.4, i.e.

$$p_x(y \mid z) = \int_\Theta f_x(y \mid \theta) \cdot \pi(\theta \mid z) \, d\theta$$

has to be adapted to the case of fuzzy data z^*.

This is possible using the integration of fuzzy valued functions described in Section 3.6. The defining equation for the predictive density becomes in the case of fuzzy a posterior density $\pi^*(\cdot \mid z^*)$:

$$p_x^*(y \mid z^*) = \oint_\Theta f_x(y \mid \theta) \odot \pi^*(\theta \mid z^*) \, d\theta \quad \forall \, y \in M_Y.$$

This generalized integration is conducted by the integration of δ-level functions, and then applying the construction lemma for characterizing functions.

The defining nested family of subsets of $\mathbb{R}$ for $p_x^*(y \mid z^*)$, i.e. $\big(A_\delta(y); \delta \in (0; 1]\big)$ with $A_\delta(y) = \big[\underline{A}_\delta(y); \overline{A}_\delta(y)\big]$ is obtained by the following equations:

$$\left. \begin{array}{l} \overline{A}_\delta(y) := \int_\Theta f_x(y \mid \theta) \cdot \overline{\pi}_\delta(\theta \mid z^*) \, d\theta \\[2mm] \text{and} \\[2mm] \underline{A}_\delta(y) := \int_\Theta f_x(y \mid \theta) \cdot \underline{\pi}_\delta(\theta \mid z^*) \, d\theta \end{array} \right\} \quad \forall \, \delta \in (0; 1].$$

From this the characterizing function $\rho(\cdot)$ of $p_x^*(y \mid z^*)$ is constructed by

$$\rho(x) := \sup \left\{ \delta \cdot I_{A_\delta(y)}(x) \colon \; \delta \in [0;1] \right\} \quad \forall\, x \in \mathbb{R}.$$

The δ-level functions $\overline{p}_{x,\delta}(\cdot \mid z^*)$ and $\underline{p}_{x,\delta}(\cdot \mid z^*)$ are determined by

$$\left.\begin{aligned}
\overline{p}_{x,\delta}(y \mid z^*) &= \overline{A}_\delta(y) \\[1mm]
\text{and} & \\[1mm]
\underline{p}_{x,\delta}(y \mid z^*) &= \underline{A}_\delta(y)
\end{aligned}\right\} \quad \forall\, \delta \in (0;1].$$

Remark 22.1: For fuzzy multivariate data $(x_i^*; y_i^*)$ the vector-characterizing functions $\zeta_i(\cdots)$ of x_i^* are frequently approximated by finitely many δ-cuts. The characterizing functions $\eta_i(\cdot)$ of y_i^* are often assumed to be of trapezoidal shape.

If the data are given in the form $(x_{i\,1}^*, \cdots, x_{i\,k}^*; y_i^*)$ usually the characterizing functions are assumed to be fuzzy intervals. Then the δ-cuts of the vector-characterizing function of the combined fuzzy sample are the Cartesian products of the δ-cuts of the individual fuzzy intervals.

22.4   Problems

(a) Let $\pi^*(\cdot)$ be a fuzzy density on $\mathbb{R}^r$ whose δ-level functions are all integrable functions with finite integral, i.e.

$$\int_{\mathbb{R}^r} \overline{\pi}_\delta(x_1, \cdots, x_r)dx_1 \cdots dx_r < \infty \qquad \forall\, \delta \in (0;1].$$

Prove that all marginal densities

$$\oint_{\mathbb{R}^{r-1}} \pi^*(x_1, \cdots, x_r)\, dx_1 \cdots dx_s\, dx_{s+1} \cdots dx_r$$

are fuzzy densities.

(b) Explain that the definition of the expectation of a fuzzy probability density $f^*(\cdot)$ specializes to the classical expectation for classical densities $f(\cdot)$ with existing expectation.

Part VII

FUZZY TIME SERIES

Many time series are the results of measurement procedures and therefore the obtained values are more or less fuzzy. Also, many economic time series contain values with remarkable uncertainty. In particular, development data and environmental time series are examples of time series whose values are not precise numbers. If such data are modeled by fuzzy numbers the resulting time series are called *fuzzy time series*.

Based on the results from Part I of this book it is necessary to consider time series with fuzzy data. In standard time series analysis the values x_t of a time series $(x_t)_{t \in T}$, where T is the series of time points, i.e. $T = \{1, 2, \cdots, N\}$, are asumed to be real numbers. By the fuzziness of many data, especially all measurement data from continuous quantities, it is necessary to consider time series whose values are fuzzy numbers x_t^*. Such time series $(x_t^*)_{t \in T}$ are called *fuzy time series*.

Mathematically a fuzzy time series is a mapping from the index set T to the set $\mathcal{F}(\mathbb{R})$ of fuzzy numbers. The generalization of time series analysis techniques to the situation of fuzzy values should fulfill the condition that it specializes to the classical techniques in the case of real valued time series.

In this part, first the necessary mathematical techniques are introduced which are used to analyze fuzzy time series. Then descriptive methods of time series analysis are generalized for fuzzy valued time series. Finally, stochastic methods of time series analysis are generalized for fuzzy time series. This Part is based on the doctoral dissertation of Hareter (2003).

23

Mathematical concepts

In this chapter mathematical concepts will be described which are used later in this part.

23.1 Support functions of fuzzy quantities

Let $\| \cdot \|$ denote the Euclidean norm, and $S^{n-1} = \{u \in \mathbb{R}^n : \|u\| = 1\}$ the unit sphere, and $\langle \cdot, \cdot \rangle$ the inner product in $\mathbb{R}^n$. Then each nonempty compact and convex subset $A \subseteq \mathbb{R}^n$ is characterized by its *support function*

$$s_A(u) := \sup\{\langle u, a \rangle : a \in A\} \qquad \forall u \in S^{n-1}.$$

An element $a \in \mathbb{R}^n$ belongs to the set A if and only if $\langle u, a \rangle \leq s_A(u)$ for all $u \in S^{n-1}$.

By a specialization of Definition 2.3 the δ-cuts of a fuzzy vector $x^* \in \mathcal{F}(\mathbb{R}^n)$ are compact and convex subsets of $\mathbb{R}^n$. Therefore x^* is characterized by

$$s_{x^*}(u, \delta) := \sup\{\langle u, a \rangle : a \in C_\delta(x^*)\} \qquad \forall u \in S^{n-1}, \forall \delta \in (0; 1] \qquad (23.1)$$

where $s_{x^*}(\cdot, \cdot)$ is called the *support function of the fuzzy vector* x^*. The set of all support functions of n-dimensional fuzzy vectors is denoted by $\mathcal{H}(\mathbb{R}^n)$, i.e.

$$\mathcal{H}(\mathbb{R}^n) := \{s_{x^*} : x^* \in \mathcal{F}(\mathbb{R}^n)\}.$$

The support function $s_{x^*}(\cdot, \cdot)$ of a fuzzy vector $x^* \in \mathcal{F}(\mathbb{R}^n)$ obeys the following [See Diamond and Kloeden (1994, page 242) and Körner (1997a, page 16)]:

Statistical Methods for Fuzzy Data   Reinhard Viertl
© 2011 John Wiley & Sons, Ltd

(1) For all $\delta \in (0; 1]$ the support function $s_{x^*}(\cdot, \delta)$ is continuous, positive homogeneous, i.e.

$$s_{x^*}(\lambda \, u, \delta) = \lambda \, s_{x^*}(u, \delta) \qquad \forall \lambda \geq 0, \ \forall u \in S^{n-1},$$

and sub-additive, i.e.

$$s_{x^*}(u_1 + u_2, \delta) \leq s_{x^*}(u_1, \delta) + s_{x^*}(u_2, \delta) \qquad \forall u_1, u_2 \in S^{n-1}.$$

(2) For all $u \in S^{n-1}$ the function $s_{x^*}(u, \cdot)$ is nonincreasing, i.e.

$$s_{x^*}(u, \alpha) \geq s_{x^*}(u, \beta) \qquad \text{for } 0 < \alpha \leq \beta \leq 1,$$

and left continuous.

Remark 23.1: For every Lebesgue integrable function $f \in L\left(S^{n-1} \times (0; 1]\right)$ fulfilling the conditions above there is exactly one fuzzy vector $x^* \in \mathcal{F}(\mathbb{R}^n)$ whose δ-cuts $C_\delta(x^*)$ are fulfilling

$$C_\delta(x^*) = \left\{ x \in \mathbb{R}^n : \langle u, x \rangle \leq f(u, \delta) \ \forall u \in \mathbb{R}^n \right\}$$

and whose support function is f [See Körner and Näther (1998, page 100)].

Example 23.1 The unit sphere in $\mathbb{R}$ is $S^0 = \{-1, 1\}$. Therefore the support function of a fuzzy number $x^* = \langle m, s, l, r \rangle_{LR}$ takes only two values:

$$s_{x^*}(u, \delta) = \begin{cases} m + s + r \, R^{(-1)}(\delta) & \text{for} \quad u = 1 \\ -m + s + l \, L^{(-1)}(\delta) & \text{for} \quad u = -1. \end{cases}$$

Later a suitable distance on the set $\mathcal{F}(\mathbb{R}^n)$ is needed. This is defined in the next subsection.

23.2   Distances of fuzzy quantities

The so-called *Hausdorff metric* $d_H(\cdot, \cdot)$ on the space $\mathcal{K}^n$ of all nonempty compact and convex subsets of $\mathbb{R}^n$ is defined for $A, B \in \mathcal{K}^n$ by

$$d_H(A, B) := \max \left\{ \sup_{a \in A} \inf_{b \in B} \|a - b\|, \ \sup_{b \in B} \inf_{a \in A} \|a - b\| \right\}.$$

With this metric the space $(\mathcal{K}^n, d_H)$ is a metric space.

There is the following relationship between this metric and the difference of support functions [see Diamond and Kloeden (1990, page 243)]:

$$d_H(A, B) = \sup_{u \in S^{n-1}} |s_A(u) - s_B(u)|. \tag{23.2}$$

Based on the Hausdorff metric different metrics can be defined on the set $\mathcal{F}(\mathbb{R}^n)$. Frequently the following distances are used:

For $x^*, y^* \in \mathcal{F}(\mathbb{R}^n)$ and $1 \le p < \infty$

$$h_p(x^*, y^*) := \left(\int_0^1 d_H\Big(C_\delta(x^*), C_\delta(y^*)\Big)^p d\delta \right)^{1/p} \tag{23.3}$$

and

$$h_\infty(x^*, y^*) := \sup_{\delta \in (0;1]} d_H\Big(C_\delta(x^*), C_\delta(y^*)\Big). \tag{23.4}$$

For details on the spaces $(\mathcal{F}(\mathbb{R}^n), h_p)$ and $(\mathcal{F}(\mathbb{R}^n), h_\infty)$ see Diamond and Kloeden (1990, 1992, 1994), Feng $et\ al.$ (2001) and Puri and Ralescu (1986). $(\mathcal{F}(\mathbb{R}^n), h_1)$ is complete and separable, $(\mathcal{F}(\mathbb{R}^n), h_\infty)$ is only complete.

Other useful metrics on $\mathcal{F}(\mathbb{R}^n)$ can be defined based on the support function $s(\cdot)$ and the metrics on the functional spaces $L^p(S^{n-1} \times (0; 1])$, $p \in [1; \infty]$. For $x^*, y^* \in \mathcal{F}(\mathbb{R}^n)$ and $1 \le p < \infty$ the metric $\rho_p(\cdot, \cdot)$ is defined by

$$\rho_p(x^*, y^*) := \|s_{x^*} - s_{y^*}\|_p = \left(n \int_0^1 \int_{S^{n-1}} |s_{x^*}(u, \delta) - s_{y^*}(u, \delta)|^p \mu(d u) d\delta \right)^{1/p} \tag{23.5}$$

where μ is denoting the Lebesgue measure on S^{n-1} with $\mu(S^{n-1}) = 1$.

For $n = 1$, i.e. $S^{n-1} = S^0 = \{-1, 1\}$, we obtain $\mu(-1) = \mu(1) = 1/2$. For $p = \infty$ the metric $\rho_\infty(\cdot, \cdot)$ is defined by

$$\rho_\infty(x^*, y^*) = \sup_{\delta \in (0;1]} \sup_{u \in S^{n-1}} |s_{x^*}(u, \delta) - s_{y^*}(u, \delta)|,$$

and from Equation (23.2) we obtain

$$\rho_\infty(x^*, y^*) = \sup_{\delta \in (0;1]} d_H\Big(C_\delta(x^*), C_\delta(y^*)\Big) = h_\infty(x^*, y^*).$$

The corresponding norm is given by

$$\|x^*\|_p := \|s_{x^*}\|_p = \left(n \int_0^1 \int_{S^{n-1}} |s_{x^*}(u, \delta)|^p \mu(d u) d\delta \right)^{1/p} \tag{23.6}$$

for $1 \le p < \infty$,
  and

$$\|x^*\|_\infty := \|s_{x^*}\|_\infty = \sup_{\delta \in (0;1]} \sup_{u \in S^{n-1}} |s_{x^*}(u, \delta)|$$

for $p = \infty$.

For more details on the spaces $\left(\mathcal{F}(\mathbb{R}^n), \rho_p\right)$ and $\left(\mathcal{F}(\mathbb{R}^n), \rho_\infty\right)$ see Diamond and Kloeden (1994) and Krätschmer (2002).

Based on the support function the space $\mathcal{F}(\mathbb{R}^n)$ can be embedded in the Banach space $(S^{n-1} \times (0; 1])$. This embedding obeys the following proposition:

Proposition 23.1: The mapping $s : x^* \mapsto s_{x^*}$ from the defining equation (23.1) is an isometry from $\mathcal{F}(\mathbb{R}^n)$ onto the positive cone $\mathcal{H}(\mathbb{R}^n) \subseteq L\left(S^{n-1} \times (0; 1]\right)$, with

$$h_p(x^*, y^*) = \begin{cases} \left(\int_0^1 \sup_{u \in S^{n-1}} |s_{x^*}(u, \delta) - s_{y^*}(u, \delta)|^p \, d\delta\right)^{1/p} & \text{for } 1 \le p < \infty \\ \sup_{\delta \in (0;1]} \sup_{u \in S^{n-1}} |s_{x^*}(u, \delta) - s_{y^*}(u, \delta)| & \text{for } p = \infty \end{cases},$$

semilinear:

$$s_{\lambda \cdot x^* \oplus \nu \cdot y^*} = \lambda \, s_{x^*} + \nu \, s_{y^*} \qquad \text{for } \lambda, \nu \ge 0 \text{ and } x^*, y^* \in \mathcal{F}(\mathbb{R}^n).$$

Moreover

$$s_{x^*}(u, \delta) \le s_{y^*}(u, \delta) \quad \Leftrightarrow \quad C_\delta(x^*) \subseteq C_\delta(y^*) \quad \forall \delta \in (0; 1].$$

For the proof see Körner (1997a, Theorem II.9, page 11) and Körner and Näther (2002, page 30).

By the foregoing proposition addition and scalar multiplication with non-negative numbers in the space $\mathcal{F}(\mathbb{R}^n)$ can be done in the functional space $L\left(S^{n-1} \times (0; 1]\right)$.

Remark 23.2: By the definition of the difference $\ominus$ and Proposition 23.1 for the support function of the difference $y^* \ominus x^*$ we obtain

$$s_{y^* \ominus x^*} = s_{y^* \oplus (-x^*)} = s_{y^*} + s_{-x^*}$$

where possibly $s_{y^* \ominus x^*}(u, \delta) \ne s_{y^*}(u, \delta) - s_{x^*}(u, \delta)$. For the support function of a linear combination $\lambda \cdot x^* \oplus \nu \cdot y^*$ of two fuzzy vectors x^* und y^*, and $\lambda, \nu \in \mathbb{R}$ we obtain

$$s_{\lambda \cdot x^* \oplus \nu \cdot y^*} = |\lambda| \, s_{\text{sign}(\lambda) \cdot x^*} + |\nu| \, s_{\text{sign}(\nu) \cdot y^*}. \tag{23.7}$$

Based on the support function different characteristics of fuzzy vectors can be obtained. The first is the so-called *Steiner point*.

Definition 23.1: The Steiner point $\sigma_{x^*} \in \mathbb{R}^n$ of a fuzzy vector x^* is given by

$$\sigma_{x^*} := n \int_0^1 \int_{S^{n-1}} u \, s_{x^*}(u, \delta) \, \mu(d\,u)\, d\,\delta.$$

Proposition 23.2: The mapping $\sigma : x^* \mapsto \sigma_{x^*}$ is linear, i.e.

$$\sigma_{\lambda \cdot x^* \oplus \nu \cdot y^*} = \lambda \, \sigma_{x^*} + \nu \, \sigma_{y^*} \qquad \text{for } \lambda, \nu \in \mathbb{R} \quad \text{and} \quad x^*, y^* \in \mathcal{F}\left(\mathbb{R}^n\right).$$

Proof: The linearity follows from (23.7) and

$$\begin{aligned}
\sigma_{-x^*} &= n \int_0^1 \int_{S^{n-1}} u \, s_{-x^*}(u, \delta) \, \mu(d\,u)\, d\,\delta \\
&= n \int_0^1 \int_{S^{n-1}} u \, \sup\{\langle u, a \rangle : a \in C_\delta(-x^*)\} \, \mu(d\,u)\, d\,\delta \\
&= n \int_0^1 \int_{S^{n-1}} (-u) \, \sup\{\langle -u, a \rangle : a \in C_\delta(-x^*)\} \, \mu\,(d\,(-u))\, d\,\delta \\
&= n \int_0^1 \int_{S^{n-1}} (-u) \, \sup\{\langle u, a \rangle : a \in C_\delta(x^*)\} \, \mu\,(d\,(-u))\, d\,\delta = -\sigma_{x^*}\,.
\end{aligned}$$

Remark 23.3: The Steiner point of a classical vector $x \in \mathbb{R}^n$ is the vector itself, i.e. $\sigma_x = x$. From this and Proposition 23.2 we obtain for the fuzzy vector $x_0^* := x^* \ominus \sigma_{x^*}$

$$\sigma_{x_0^*} = \sigma_{x^* \ominus \sigma_{x^*}} = \sigma_{x^*} - \sigma_{\sigma_{x^*}} = 0.$$

Example 23.2 For a fuzzy number $x^* = \langle m, s, l, r \rangle_{LR}$ of LR-type the Steiner point is given by

$$\sigma_{x^*} = m + r\,\sigma_{x_R^*} - l\,\sigma_{x_L^*}.$$

Proposition 23.3: The Steiner point σ_{x^*} of a fuzzy vector x^* fulfills the following:

$$\min_{a \in \mathbb{R}^n} \rho_2(a, x^*) = \rho_2(\sigma_{x^*}, x^*).$$

For $x_0^* = x^* \ominus \sigma_{x^*}$ and $y_0^* = y^* \ominus \sigma_{y^*}$ we have

$$\rho_2^2(x^*, y^*) = \rho_2^2(x_0^*, y_0^*) + \|\sigma_{x^*} - \sigma_{y^*}\|_2^2. \tag{23.8}$$

The proof is given in Körner (1997a, Theorem II.16, page 17).

Remark 23.4: For $p = 2$ the space $L^p\left(S^{n-1} \times (0; 1]\right)$ is a Hilbert space. Therefore in the following the metric $\rho_2(\cdot, \cdot)$ on $\mathcal{F}(\mathbb{R}^n)$ is used frequently. Moreover only fuzzy vectors in $\mathcal{F}_2\left(\mathbb{R}^n\right)$, defined by

$$\mathcal{F}_2\left(\mathbb{R}^n\right) := \left\{x^* \in \mathcal{F}\left(\mathbb{R}^n\right) : \|x^*\|_2 < \infty\right\},$$

will be considered in order to guarantee the existence of the metric $\rho_2(\cdot, \cdot)$. The space of all support functions $s_{x^*}(\cdot, \cdot)$ with $x^* \in \mathcal{F}_2\left(\mathbb{R}^n\right)$ will be denoted by $\mathcal{H}_2(\mathbb{R}^n)$.

Proposition 23.4: The space $\left(\mathcal{F}_2\left(\mathbb{R}^n\right), \rho_2\right)$ is isomorphic with the closed and convex cone $\mathcal{H}_2(\mathbb{R}^n)$ of the Hilbert space $L^2\left(S^{n-1} \times (0; 1]\right)$ with the inner product

$$\langle x^*, y^* \rangle := \langle s_{x^*}, s_{y^*} \rangle = n \int_0^1 \int_{S^{n-1}} s_{x^*}(u, \delta)\, s_{y^*}(u, \delta)\, \mu(d\,u)\, d\,\delta. \tag{23.9}$$

For the proof see Körner (1997a, page 15).

The closedness of the cone which is isomorphic to $\left(\mathcal{F}_2\left(\mathbb{R}^n\right), \rho_2\right)$ is important for least sum of squares approximations in linear models.

Proposition 23.5: The inner product defined by (23.9), for $x^*, y^*, z^* \in \mathcal{F}_2\left(\mathbb{R}^n\right)$ and $\lambda \in \mathbb{R}_0^+$ obeys the following:

(1) $\langle x^*, x^* \rangle \geq 0$ and $\langle x^*, x^* \rangle = 0 \Leftrightarrow x^* = \mathbf{0}$.

(2) $\langle x^*, y^* \rangle = \langle y^*, x^* \rangle$.

(3) $\langle x^* \oplus y^*, z^* \rangle = \langle x^*, z^* \rangle + \langle y^*, z^* \rangle$.

(4) $\langle \lambda \cdot x^*, y^* \rangle = \lambda \langle x^*, y^* \rangle$.

(5) $|\langle x^*, y^* \rangle| \leq \sqrt{\langle x^*, x^* \rangle \langle y^*, y^* \rangle}$.

Moreover

$$\rho_2(x^*, y^*) = \sqrt{\langle x^*, x^* \rangle - 2 \langle x^*, y^* \rangle + \langle y^*, y^* \rangle}. \tag{23.10}$$

The proof follows from the definition of the inner product; see also Feng (2001, page 12).

Similar to real vectors we have the following relationships for fuzzy vectors:

$$\langle x^*, y^* \rangle = \frac{1}{2}\left(\|x^* \oplus y^*\|_2^2 - \|x^*\|_2^2 - \|y^*\|_2^2\right) \quad \text{and} \quad \|x^*\|_2^2 = \langle x^*, x^* \rangle,$$

which is seen by application of the definition of the inner product (23.9) and the norm (23.6).

Example 23.3 For the inner product $\langle x^*, y^* \rangle$ of two fuzzy numbers $x^* = \langle m_x, s_x, l_x, r_x \rangle_{LR}$ and $y^* = \langle m_y, s_y, l_y, r_y \rangle_{LR}$ of LR-form we obtain:

$$\langle x^*, y^* \rangle = m_x\, m_y + s_x\, s_y + \|x_L^*\|_2^2\, l_x\, l_y + \|x_R^*\|_2^2\, r_x\, r_y$$
$$+ \|x_L^*\|_1 \left(-m_x\, l_y - m_y\, l_x + s_x\, l_y + s_y\, l_x \right)$$
$$+ \|x_R^*\|_1 \left(m_x\, r_y + m_y\, r_x + s_x\, r_y + s_y\, r_x \right)$$

where

$$\|x_L^*\|_2^2 = \frac{1}{2} \int_0^1 \left(L^{(-1)}(\delta) \right)^2 d\,\delta, \qquad \|x_L^*\|_1 = \frac{1}{2} \int_0^1 L^{(-1)}(\delta)\, d\,\delta$$

and $\|x_R^*\|_2^2$ and $\|x_R^*\|_1$ are defined analogously. For the distance $\rho_2(x^*, y^*)$ of these fuzzy numbers we obtain:

$$\rho_2^2(x^*, y^*) = (m_x - m_y)^2 + (s_x - s_y)^2 + \|x_L^*\|_2^2 (l_x - l_y)^2 + \|x_R^*\|_2^2 (r_x - r_y)^2$$
$$- 2\, \|x_L^*\|_1 \left[(m_x - m_y) - (s_x - s_y) \right] (l_x - l_y)$$
$$+ 2\, \|x_R^*\|_1 \left[(m_x - m_y) + (s_x - s_y) \right] (r_x - r_y).$$

In the case of trapezoidal fuzzy numbers $t^*(m, s, l, r)$ we have

$$L^{(-1)}(\delta) = R^{(-1)}(\delta) = 1 - \delta$$

and

$$\|x_L^*\|_1 = \|x_R^*\|_1 = \frac{1}{2} \int_0^1 (1 - \delta)\, d\,\delta = \frac{1}{4}$$

$$\|x_L^*\|_2^2 = \|x_R^*\|_2^2 = \frac{1}{2} \int_0^1 (1 - \delta)^2\, d\,\delta = \frac{1}{6}\,.$$

23.3 Generalized Hukuhara difference

The generalized difference operation $\ominus$ for fuzzy numbers via the extension principle increases the fuzziness. For symmetric fuzzy numbers x^* in LR-form with $L(\cdot) = R(\cdot)$, $l = r$ and $m = 0$ we obtain $x^* \ominus x^* = x^* \oplus (-x^*) \neq 0$. In general for given $x^* \oplus y^*$ and given x^* the vector y^* cannot be calculated by $(x^* \oplus y^*) \ominus x^*$. Moreover the difference $\ominus$ does not correspond to the difference in the functional space $L^2 \left(S^{n-1} \times (0; 1] \right)$, i.e. the support function of $y^* \ominus x^*$ is not equal to $s_{y^*} - s_{x^*}$.

In order to find a difference which corresponds more to the usual difference of numbers the following argument is helpful: The difference $x - y$ of two real numbers x and y is the real number z for which $y + z = x$. This results in the so-called *Hukuhara difference*.

Definition 23.2: For $x^* \in \mathcal{F}_2(\mathbb{R}^n)$ and $y^* \in \mathcal{F}_2(\mathbb{R}^n)$ the so-called Hukuhara difference $y^* \ominus_H x^*$ is – if it exists – the solution $h^* \in \mathcal{F}_2(\mathbb{R}^n)$ of the equation $x^* \oplus h^* = y^*$

Remark 23.5: In Diamond and Kloeden (1994) the following is proved. If the Hukuhara difference of two fuzzy numbers x^* and y^* exists in $L^2\left(S^{n-1} \times (0; 1]\right)$, then for the corresponding support functions it follows $s_{y^* \ominus_H x^*} = s_{y^*} - s_{x^*}$ and the δ-cuts of $y^* \ominus_H x^*$ are given by

$$C_\delta(y^* \ominus_H x^*) = \left\{z \in \mathbb{R}^n : C_\delta(x^*) + z \subseteq C_\delta(y^*)\right\} \qquad \forall \delta \in (0; 1].$$

The Hukuhara difference need not exist for all x^*, $y^* \in \mathcal{F}_2(\mathbb{R}^n)$. If $\mathrm{supp}(y^*) \subset \mathrm{supp}(x^*)$ no element $h^* \in \mathcal{F}_2(\mathbb{R}^n)$ is possible obeying $x^* \oplus h^* = y^*$.

Example 23.4 The Hukuhara difference $y^* \ominus_H x^*$ of two symmetric triangular fuzzy numbers $x^* = d^*(m_x, l_x, l_x)$ and $y^* = d^*(m_y, l_y, l_y)$ with $l_y < l_x$ does not exist. In the case of $l_y \geq l_x$ the Hukuhara difference of $y^* \ominus_H x^*$ is the triangular fuzzy number $d^*(m_y - m_x, l_y - l_x, l_y - l_x)$.

By the above results it is necessary to construct a more general difference for two fuzzy vectors. This is possible by a generalization of the Hukuhara difference.

Definition 23.3: For $x^*, y^* \in \mathcal{F}_2(\mathbb{R}^n)$ the *generalized Hukuhara difference* $\sim_H$ is the l^2-approximation of the Hukuhara difference, i.e.

$$y^* \sim_H x^* := h^* \quad \Leftrightarrow \quad \rho_2(x^* \oplus h^*, y^*) = \min_{z^* \in \mathcal{F}_2(\mathbb{R}^n)} \rho_2(x^* \oplus z^*, y^*). \quad (23.11)$$

The generalized Hukuhara difference $\sim_H$ is an extension of the Hukuhara difference $\ominus_H$ to the whole space $\mathcal{F}_2(\mathbb{R}^n)$, i.e. if $y^* \ominus_H x^*$ exists it is identical to $y^* \sim_H x^*$.

Remark 23.6: For the generalized Hukuhara difference usually

$$(x^* \sim_H y^*) \oplus y^* \neq x^*,$$

i.e. the generalized Hukuhara difference is not a difference in $\mathcal{F}_2(\mathbb{R}^n)$ in the standard sense. The inequality holds if y^* is more fuzzy than x^*. In this situation the result $x^* \sim_H y^*$ is a real number. But for known sum $x^* \oplus y^*$ and given x^* the generalized Hukuhara difference $(x^* \oplus y^*) \sim_H x^*$ is the best possible value for the unknown y^*.

Contrary to the operation $\ominus$ for the operations $\ominus_H$ and $\sim_H$ the fuzziness of the result is less than the fuzziness of the minuend, moreover by $x^* \oplus 0 = x^*$ we have $x^* \ominus_H x^* = x^* \sim_H x^* = 0$.

Remark 23.7: For real numbers we have $y \ominus x = y \sim_H x = y - x$. For fuzzy numbers and fuzzy vectors and given sum and one quantity the other quantity is best calculated by the generalized Hukuhara difference. The existence and uniqueness of $y^* \sim_H x^*$ follows from the projection theorem in Hilbert spaces.

Theorem 23.1: Let $(H, \|\cdot\|)$ be a Hilbert space, $x \in H$ and $K \neq \emptyset$ a closed convex subset of H. Then there exists exactly one element $y \in K$ obeying

$$\|x - y\| = \inf_{z \in K} \|x - z\|.$$

The proof is given in books on functional analysis.

By Proposition 23.4 the space $(\mathcal{F}_2(\mathbb{R}^n), \rho_2)$ is isomorphic to the closed convex cone $\mathcal{H}_2(\mathbb{R}^n)$ in the Hilbert space $L^2(S^{n-1} \times (0;1])$. The approximation problem (23.11) for the calculation of the generalized Hukuhara difference of two fuzzy vectors $x^*, y^* \in \mathcal{F}_2(\mathbb{R}^n)$ can be transformed to an equivalent minimization problem in $L^2(S^{n-1} \times (0;1])$. This equivalent problem is

$$\min_{s_{z^*} \in \mathcal{H}_2(\mathbb{R}^n)} \|(s_{x^*} + s_{z^*}) - s_{y^*}\|_2 = \min_{s_{z^*} \in \mathcal{H}_2(\mathbb{R}^n)} \|(s_{y^*} - s_{x^*}) - s_{z^*}\|_2.$$

The support function of the generalized Hukuhara difference $s_{y^* \sim_H x^*}$ is the unique orthogonal projection of the function $s_{y^*} - s_{x^*}$ onto the cone of support functions $\mathcal{H}_2(\mathbb{R}^n)$.

Another important property of the generalized Hukuhara difference $\sim_H$ is the following:

$$\sigma_{y^* \sim_H x^*} = \sigma_{y^*} - \sigma_{x^*}$$

In general the calculation of the generalized Hukuhara difference can be complicated. For some special cases the result can be given in closed form. An important tool for this is the Kuhn–Tucker theorem.

Theorem 23.2: Kuhn–Tucker
Let C be a symmetric positive semidefinite matrix. Then the solution of the quadratic optimization problem

$$\min_{Ax \le a} (x - c)^T C (x - c)$$

is fulfilling the so-called Kuhn–Tucker condition

$$2C(x - c) + A^T u = 0, \qquad Ax + y = a,$$
$$u \ge 0, \quad y \ge 0 \quad \text{and} \quad u^T y = 0.$$

The proof is given in Luptáčik (1981, Theorem IV.C, page 118).

A special case where the solution of the generalized Hukuhara difference can be given in closed form is the generalized Hukuhara difference of triangular fuzzy numbers.

The generalized Hukuhara difference of two trapezoidal fuzzy numbers is not possible in a simple way. An example is the following.

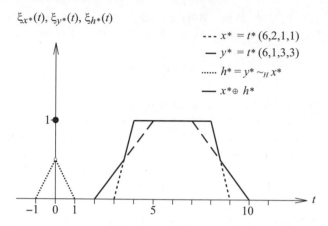

$\xi_{x^*}(t), \xi_{y^*}(t), \xi_{h^*}(t)$

--- $x^* = t^*(6,2,1,1)$
— $y^* = t^*(6,1,3,3)$
...... $h^* = y^* \sim_H x^*$
— $x^* \oplus h^*$

Figure 23.1 Generalized Hukuhara difference of two trapezoidal fuzzy numbers.

Example 23.5 Let x^* and y^* be two trapezoidal fuzzy numbers $x^* = t^*(6, 2, 1, 1)$ and $y^* = t^*(6, 1, 3, 3)$. Then $h^* = y^* \sim_H x^*$ has a characterizing function of polygonal type (cf. Figure 23.1).

In many cases the generalized Hukuhara difference $y^* \sim_H x^*$ of two trapezoidal fuzzy numbers $x^* = t(m_x, s_x, l_x, r_x)$ and $y^* = t(m_y, s_y, l_y, r_y)$ is again a trapezoidal fuzzy number. For $s_x > s_y$ and $(2\,s_x + l_x + r_x) < (2\,s_y + l_y + r_y)$ as in the above Example 23.3, i.e. the support of x^* is less fuzzy than the support of y^*, and the 1-cut of x^* a larger uncertainty, then the result is a polygonal fuzzy number.

$$h^* = \{(x_1, 0), (x_2, y_2), (x_3, 1), (x_4, 1), (x_5, y_5), (x_6, 0)\}.$$

The determination of the numbers $x_1, x_2, x_3, x_4, x_5, x_6, y_2$ and y_5 is done by

$$\min_{x_1,x_2,x_3,x_4,x_5,x_6,y_2,y_5} \rho_2(x^* \oplus h^*, y^*)$$

under the conditions $x_i \le x_{i+1}$, $i = 1\,(1)\,5$, and $0 \le y_2, y_5 \le 1$.

The distance is calculated using the abbreviations $f_1 = m_x - m_y - s_x + s_y$ and $f_2 = m_x - m_y + s_x - s_y$ in the following way:

$$\rho_2^2(x^* \oplus h^*, y^*) = \int_0^{y_2} \left(f_1 - (1 - \delta)(l_x - l_y) + x_1 + \delta \frac{x_2 - x_1}{y_2} \right)^2 d\delta$$

$$+ \int_{y_2}^1 \left(f_1 + x_3 + (1 - \delta)\left(l_y - l_x + \frac{x_2 - x_3}{1 - y_2} \right) \right)^2 d\delta$$

$$+ \int_0^{y_5} \left(f_2 + (1 - \delta)(r_x - r_y) + x_6 + \delta\, \frac{x_5 - x_6}{y_5} \right)^2 d\delta$$

$$+ \int_{y_5}^1 \left(f_2 + x_4 + (1 - \delta)\left(r_x - r_y + \frac{x_5 - x_4}{1 - y_5} \right) \right)^2 d\delta$$

For this minimization problem the methods of nonlinear programming can be used. For details see Luptáčik (1981).

The Hukuhara difference of two polygonal fuzzy numbers is again of polygonal type. The calculation complexity increases exponentially with the number of points describing the polygons.

In applications the complexity of the calculation can be reduced by looking only to a limited number of δ-cuts.

Another possibility to reduce the complexity is to restrict the solution space for the projection.

Remark 23.8: If the solution space for the Hukuhara difference $h^* = y^* \sim_H x^*$ of two trapezoidal fuzzy numbers $x^* = t^*(m_x, s_x, l_x, r_x)$ and $y^* = t^*(m_y, s_y, l_y, r_y)$ is restricted to the set of trapezoidal fuzzy numbers, the parameters m_h, s_h, l_h and r_h of the solution $h^* = t^*(m_h, s_h, l_h, r_h)$ with $\widehat{s}_h := s_y - s_x$, $\widehat{l}_h := l_y - l_x$, $\widehat{r}_h := r_y - r_x$, $r_1 := \|x_R^*\|_1$, $l_1 := \|x_L^*\|_1$, $r_2 := \|x_R^*\|_2^2$ and $l_2 := \|x_L^*\|_2^2$ can be calculated in the following way:

$$
s_h =
\begin{cases}
\max\{\widehat{s}_h, 0\} & \text{for} \quad \widehat{l}_h, \widehat{r}_h \geq 0 \\[2mm]
\max\{\widehat{s}_h + r_1 \widehat{r}_h, 0\} & \text{for} \quad \widehat{l}_h \geq 0, \widehat{r}_h < 0 \\[2mm]
\max\{\widehat{s}_h + l_1 \widehat{l}_h, 0\} & \text{for} \quad \widehat{l}_h < 0, \widehat{r}_h \geq 0 \\[2mm]
\max\{\widehat{s}_h + l_1 \widehat{l}_h + r_1 \widehat{r}_h, 0\} & \text{for} \quad \widehat{l}_h, \widehat{r}_h < 0
\end{cases}
$$

$$
l_h =
\begin{cases}
\max\{\widehat{l}_h, 0\} & \text{for} \quad \widehat{s}_h, \widehat{r}_h \geq 0 \\[3mm]
\max\left\{\widehat{l}_h + \dfrac{l_1 r_1}{l_2 - l_1^2} \widehat{r}_h, 0\right\} & \text{for} \quad \widehat{s}_h \geq 0, \widehat{r}_h < 0 \\[3mm]
\max\left\{\widehat{l}_h + \dfrac{l_1 (r_2 - 2 r_1^2)}{(r_2 - r_1^2)(l_2 - l_1^2) - l_1^2 r_1^2} \widehat{s}_h, 0\right\} & \text{for} \quad \widehat{s}_h < 0, \widehat{r}_h \geq 0 \\[3mm]
\max\left\{\widehat{l}_h + \dfrac{l_1 \widehat{s}_h + l_1 r_1 \widehat{r}_h}{l_2 - l_1^2}, 0\right\} & \text{for} \quad \widehat{s}_h, \widehat{r}_h < 0
\end{cases}
$$

$$r_h = \begin{cases} \max\{\widehat{r}_h, 0\} & \text{for} \quad \widehat{s}_h, \widehat{l}_h \geq 0 \\[2ex] \max\left\{\widehat{r}_h + \dfrac{l_1 r_1}{r_2 - r_1^2}\widehat{l}_h, 0\right\} & \text{for} \quad \widehat{s}_h \geq 0, \widehat{l}_h < 0 \\[2ex] \max\left\{\widehat{r}_h + \dfrac{r_1(l_2 - 2l_1^2)}{(r_2 - r_1^2)(l_2 - l_1^2) - l_1^2 r_1^2}\widehat{s}_h, 0\right\} & \text{for} \quad \widehat{s}_h < 0, \widehat{l}_h \geq 0 \\[2ex] \max\left\{\widehat{r}_h + \dfrac{r_1\widehat{s}_h + l_1 r_1 \widehat{l}_h}{r_2 - r_1^2}, 0\right\} & \text{for} \quad \widehat{s}_h, \widehat{l}_h < 0 \end{cases}$$

$$m_h = m_y - m_x + r_1\left(\widehat{r}_h - r_h\right) - l_1\left(\widehat{l}_h - l_h\right).$$

The proof is given in Munk (1988, page 33).

For the examples in the following chapter the solution space for the calculations of Hukuhara differences of two trapezoidal fuzzy numbers is restricted to the set of trapezoidal fuzzy numbers, and the calculations of Example 23.3 are used.

24

Descriptive methods for fuzzy time series

Descriptive methods of time series analysis work without stochastic models. The goal of these methods is to find trends and seasonal influence in time series by elementary methods. A short survey of such methods for standard time series is given in Janacek (2001).

A fuzzy time series $(x_t^*)_{t \in T}$ is an ordered sequence of fuzzy numbers, where usually $T = \{1, 2, \dots, N\}$. Formally a one-dimensional fuzzy time series is a mapping $T \to \mathcal{F}(\mathbb{R})$ which gives for any time point t a fuzzy number x_t^*. Classical time series are special forms of fuzzy time series.

A reasonable generalization should generate the classical results in the case of classical data. This is guaranteed if the extension principle is used.

For the approximation by a polynomial trend in the case of fuzzy time series see Körner (1997a), Körner and Näther (1998), Munk (1998), Näther and Albrecht (1990).

24.1 Moving averages

In order to obtain an overview concerning the long time behavior of a time series it is useful to eliminate the random oscillations of an observed time series $(x_t^*)_{t \in T}$. This elimination is possible by local approximation. A simple way to do this is by smoothing the values of the time series by a local arithmetic mean. This smoothing can be taken from classical time series analysis, i.e.

$$y_t = \frac{1}{2q + 1} \sum_{i=-q}^{q} x_{t+i}, \qquad t = q + 1\,(1)\,N - q\,,$$

and application of the extension principle in the case of fuzzy data x_t^*. For that, first the fuzzy numbers $x_{t-q}^*, \ldots, x_{t+q}^*$ have to be combined to a fuzzy vector $\boldsymbol{x}^* \in \mathcal{F}\left(\mathbb{R}^{2q+1}\right)$ with vector-characterizing function $\xi_{\boldsymbol{x}^*}(\cdot, \ldots, \cdot)$. From this the fuzzy value y_t^* and its characterizing function $\xi_{y_t^*}(\cdot)$ is given by its values $\xi_{y_t^*}(y)$ for all $y \in \mathbb{R}$ by

$$\xi_{y_t^*}(y) = \sup \left\{ \xi_{\boldsymbol{x}^*}(x_{-q}, \ldots, x_q) : (x_{-q}, \ldots, x_q) \in \mathbb{R}^{2q+1} : \frac{1}{2q+1} \sum_{i=-q}^{q} x_i = y \right\}.$$

The δ-cuts $C_\delta(y_t^*)$ are given by theorem 3.1

$$C_\delta(y_t^*) = \left[\min_{(x_{-q}, \ldots, x_q) \in C_\delta(\boldsymbol{x}^*)} \frac{1}{2q+1} \sum_{i=-q}^{q} x_i, \max_{(x_{-q}, \ldots, x_q) \in C_\delta(\boldsymbol{x}^*)} \frac{1}{2q+1} \sum_{i=-q}^{q} x_i \right].$$

The result can be given in short by

$$y_t^* = \frac{1}{2q+1} \cdot \bigoplus_{i=-q}^{q} x_{t+i}^*, \qquad t = q+1\,(1)\,N - q,$$

For fuzzy observations $(x_t^*)_{t \in T}$ with approximately linear behavior, i.e. $x_t^* \approx a^* \oplus t \cdot b^*$ and corresponding δ-cuts of the coefficients $C_\delta(a^*) = [\underline{a}_\delta, \overline{a}_\delta]$ and $C_\delta(b^*) = [\underline{b}_\delta, \overline{b}_\delta]$, respectively, the δ-cuts $C_\delta(x_t^*)$ can be written, using random variables $\underline{\varepsilon}_{t,\delta}$ and $\overline{\varepsilon}_{t,\delta}$ in the following way:

$$C_\delta(x_t^*) = \left[\underline{a}_\delta + \underline{b}_\delta\, t + \underline{\varepsilon}_{t,\delta}, \overline{a}_\delta + \overline{b}_\delta\, t + \overline{\varepsilon}_{t,\delta}\right].$$

The random variables $(\underline{\varepsilon}_{t,\delta})_{t \in T}$ and $(\overline{\varepsilon}_{t,\delta})_{t \in T}$ can be considered as independent relative to the time, obeying $\mathbb{E}\,\underline{\varepsilon}_{t,\delta} = 0$ and $\mathbb{E}\,\overline{\varepsilon}_{t,\delta} = 0$ with finite variances Var $\underline{\varepsilon}_{t,\delta}$ and Var $\overline{\varepsilon}_{t,\delta}$. For fixed time $t \in T$ the random variables $\underline{\varepsilon}_{t,\delta}$ and $\overline{\varepsilon}_{t,\delta}$ are dependent. For the δ-cuts $C_\delta(y_t^*)$ of the mean value y_t^* we obtain

$$C_\delta(y_t^*) = \left[\min_{(x_{-q}, \ldots, x_q) \in C_\delta(\boldsymbol{x}^*)} \frac{1}{2q+1} \sum_{i=-q}^{q} x_i, \max_{(x_{-q}, \ldots, x_q) \in C_\delta(\boldsymbol{x}^*)} \frac{1}{2q+1} \sum_{i=-q}^{q} x_i \right]$$

$$= \left[\frac{1}{2q+1} \sum_{i=-q}^{q} (\underline{a}_\delta + \underline{b}_\delta\,(t+i) + \underline{\varepsilon}_{t+i,\delta}), \frac{1}{2q+1} \sum_{i=-q}^{q} (\overline{a}_\delta + \overline{b}_\delta\,(t+i) + \overline{\varepsilon}_{t+i,\delta}) \right]$$

$$= \left[\underline{a}_\delta + \underline{b}_\delta\, t + \frac{1}{2q+1} \sum_{i=-q}^{q} \underline{\varepsilon}_{t+i,\delta}, \overline{a}_\delta + \overline{b}_\delta\, t + \frac{1}{2q+1} \sum_{i=-q}^{q} \overline{\varepsilon}_{t+i,\delta} \right]$$

$$\approx \left[\underline{a}_\delta + \underline{b}_\delta\, t + \mathbb{E}\,\underline{\varepsilon}_{t,\delta}, \overline{a}_\delta + \overline{b}_\delta\, t + \mathbb{E}\,\overline{\varepsilon}_{t,\delta}\right] = \left[\underline{a}_\delta + \underline{b}_\delta\, t, \overline{a}_\delta + \overline{b}_\delta\, t\right].$$

x_t^*, y_t^*

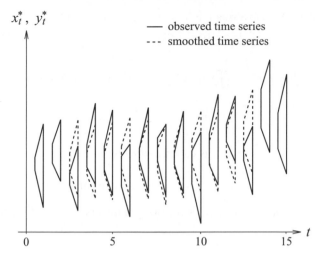

Figure 24.1   Moving averages of length 2.

For strict linear time series, like in classical time series analysis, the moving averages are identical to the original time series.

Moving average procedures are a kind of *filtering* of time series. The filtered time series is smoother if more observations are used for the filtration, i.e. as larger q is. But on the boundaries of T it is not possible to obtain filtered values. The smoothed time series (y_t^*) is shorter than the original time series $(x_t^*)_{t \in T}$.

Example 24.1 Figure 24.1 gives a fuzzy time series with moving averages of length 2. Here $T = \{1, 2, \cdots, 15\}$. The solid trapezoids are the characterizing functions of the original values $\left(x_t^*\right)_{t=1\,(1)\,15} = (t^*(m_t, s_t, l_t, r_t))_{t=1\,(1)\,15}$. The support of the individual measurements is the basis of the trapezoid. The characterizing functions of the values of the smoothed time series $(y_t^*)_{t=3(1)13}$ are depicted by broken lines.

24.2   Filtering

Filtering or filtration of time series are procedures to transform the values of a time series. Moving averages are special forms of filtering procedures.

24.2.1   Linear filtering

Moving averages are examples of linear transformations of fuzzy time series $(x_t^*)_{t \in T}$ to a smoothed time series (y_t^*). A general method from classical time series analysis can be adapted for fuzzy time series. This generalization is given in the following definition.

Definition 24.1: A transformation $L = (a_{-q}, \ldots, a_s) \in \mathbb{R}^{s+q+1}$ of a time series $(x_t^*)_{t \in T}$ to a time series $(y_t^*)_{t=q+1\,(1)\,N-s}$ by

$$y_t^* = L\,x_t^* = \bigoplus_{i=-q}^{s} a_i \cdot x_{t+i}^*, \qquad t = q+1\,(1)\,N - s,$$

is called a *linear filter*.

A linear filter L with $\displaystyle\sum_{i=-q}^{s} a_i = 1$ is called a *moving average*.

If $s = q$ with $a_i = (2q+1)^{-1}$, $i = -q\,(1)\,q$ it is called an *ordinary moving average*.

Remark 24.1: By the following rules for fuzzy numbers $x^*, y^*, z^* \in \mathcal{F}(\mathbb{R})$ and real numbers $\alpha, \beta, \gamma \in \mathbb{R}$,

$$\alpha \cdot (\beta \cdot x^* \oplus \gamma \cdot y^*) = (\alpha\,\beta) \cdot x^* \oplus (\alpha\,\gamma) \cdot y^*$$

and

$$(\alpha \cdot x^* \oplus \beta \cdot y^*) \oplus \gamma \cdot z^* = \alpha \cdot x^* \oplus (\beta \cdot y^* \oplus \gamma \cdot z^*)$$

for two fuzzy time series $(x_t^*)_{t \in T}$ and $(y_t^*)_{t \in T}$, and $\alpha, \beta \in \mathbb{R}$ we obtain the following:

$$L\,(\alpha \cdot x_t^* \oplus \beta \cdot y_t^*) =$$

$$= \bigoplus_{i=-q}^{s} a_i \cdot (\alpha \cdot x_{t+i}^* \oplus \beta \cdot y_{t+i}^*) = \bigoplus_{i=-q}^{s} \left((a_i\,\alpha) \cdot x_{t+i}^* \oplus (a_i\,\beta) \cdot y_{t+i}^* \right)$$

$$= \left(\bigoplus_{i=-q}^{s} (a_i\,\alpha) \cdot x_{t+i}^* \right) \oplus \left(\bigoplus_{i=-q}^{s} (a_i\,\beta) \cdot y_{t+i}^* \right) = \alpha \cdot L\,x_t^* \oplus \beta \cdot L\,y_t^* \quad (24.1)$$

Proposition 24.1: The δ-cuts $C_\delta(y_t^*)$ of the smoothed time series $y_t^* = L\,x_t^*$ can be calculated from the δ-cuts $C_\delta(x_t^*) = \left[\underline{x}_{t,\delta}, \overline{x}_{t,\delta}\right]$ of the observed time series $(x_t^*)_{t \in T}$ by

$$C_\delta(y_t^*) = C_\delta \left(\bigoplus_{i=-q}^{s} a_i \cdot x_{t+i}^* \right) = \left[\sum_{i=-q}^{s} a_i\,\underline{c}_{t+i}, \; \sum_{i=-q}^{s} a_i\,\overline{c}_{t+i} \right]$$

with

$$\underline{c}_{t+i} = \begin{cases} \underline{x}_{t+i,\delta} & \text{for} \quad a_i \geq 0 \\ \overline{x}_{t+i,\delta} & \text{for} \quad a_i < 0 \end{cases} \quad \text{and} \quad \overline{c}_{t+i} = \begin{cases} \overline{x}_{t+i,\delta} & \text{for} \quad a_i \geq 0 \\ \underline{x}_{t+i,\delta} & \text{for} \quad a_i < 0. \end{cases}$$

Proof: Applying Theorem 3.1 with $f(x_{-q}, \ldots, x_s) = \sum_{i=-q}^{s} a_i x_i$, by

$$\min_{x \in C_\delta(x^*)} a x = \begin{cases} a\, \underline{x}_\delta & \text{for} \quad a \geq 0 \\ a\, \overline{x}_\delta & \text{for} \quad a < 0 \end{cases} \quad \text{and} \quad \max_{x \in C_\delta(x^*)} a x = \begin{cases} a\, \overline{x}_\delta & \text{for} \quad a \geq 0 \\ a\, \underline{x}_\delta & \text{for} \quad a < 0 \end{cases}$$

the result is obtained.

Remark 24.2: If negative coefficients are used in moving averages the fuzziness of the smoothed time series $(y_t^*)_{t=q+1\,(1)\,N-s}$ becomes larger than the observed time series $(x_t^*)_{t \in T}$. For example for constant observations $x_t^* = x^*$, $t = 1\,(1)\,N$, the smoothed values y_t^* for $t = q + 1\,(1)\,N - s$ is not the original constant time series. For

$$y_t^* = \bigoplus_{i=-q}^{s} a_i \cdot x_{t+i}^* = \bigoplus_{i=-q}^{s} a_i \cdot x^* \quad \text{with} \quad \sum_{i=-q}^{s} a_i = 1 \,,$$

with $a_i > 0, i = -q\,(1)\,s - 1, a_s < 0$ and $C_\delta(x_t^*) = \left[\underline{x}_{t,\delta}, \overline{x}_{t,\delta}\right] = \left[\underline{x}_\delta, \overline{x}_\delta\right], t = 1\,(1)\,N$, by Proposition 24.1 for the upper limit $\overline{y}_{t,\delta}$ of the δ-cut $C_\delta(y_t^*) = \left[\underline{y}_{t,\delta}, \overline{y}_{t,\delta}\right]$ we obtain

$$\overline{y}_{t,\delta} = \sum_{i=-q}^{s-1} a_i\, \overline{x}_\delta + a_s\, \underline{x}_\delta \geq \left(\sum_{i=-q}^{s-1} a_i\right) \overline{x}_\delta + a_s\, \overline{x}_\delta = \overline{x}_\delta.$$

The values $(y_t^*)_{t=q+1\,(1)\,N-s}$ for $\underline{x}_\delta < \overline{x}_\delta$ are more fuzzy than $(x_t^*)_{t \in T}$. For the same reason filters with fuzzy coefficients a_i^* seem to be unreasonable.

The construction of filters poses the following question: Is there a nontrivial filter $L = (a_{-2}, a_{-1}, a_0, a_1, a_2) \neq (0, 0, 1, 0, 0)$ which results in the original time series for all observations x_t^* with quadratic form $x_t^* = b_1^* \oplus b_2^* \cdot t \oplus b_3^* \cdot t^2, t \in T$ with $b_1^*, b_2^*, b_3^* \in \mathcal{F}(\mathbb{R})$, i.e. $y_t^* = L\,x_t^*$.

For classical real valued observations $(x_t)_{t \in T}$ the filter $L = (a_{-2}, a_{-1}, a_0, a_1, a_2) = 35^{-1}(-3, 12, 17, 12, -3)$ is fulfilling this; see Schlittgen and Streitberg (1984, page 36) or Janacek (2001, page 13). For fuzzy data a filter with the desired properties can have positive coefficients only. For the following reasons the construction of such a filter for fuzzy time series is not possible. For this look at the classical time series $x_t = b_1 + b_2\, t + b_3\, t^2$.

The condition in this case is

$$\sum_{i=-2}^{2} a_i \left(b_1 + b_2\,(t-i) + b_3\,(t-i)^2\right) = b_1 + b_2\, t + b_3\, t^2 \qquad \text{for } t = 2\,(1)\,N - 2 \,.$$

From that, for the determination of the coefficients a_i we obtain the following equations:

$$a_{-2} + a_{-1} + a_0 + a_1 + \quad a_2 = 1$$
$$-2\,a_{-2} - a_{-1} + 0 + a_1 + 2\,a_2 = 0$$
$$4\,a_{-2} + a_{-1} + 0 + a_1 + 4\,a_2 = 0$$

By $a_i \geq 0, i = -2\,(1)\,2$ from the third equation we obtain $a_{-2} = a_{-1} = a_1 = a_2 = 0$, and from the first equation $a_0 = 1$, i.e. the equations can only be fulfilled for the trivial filter. By the same considerations it follows: Also for polynomial fuzzy time series $(x_t^*)_{t \in T}$ obeying $x_t^* = b_1^* \oplus b_2^* \cdot t \oplus \ldots \oplus b_n^* \cdot t^n, n \geq 2$ with $b_i^* \in \mathcal{F}(\mathbb{R}), i = 1\,(1)\,n$, there is no nontrivial filter with positive coefficients, such that the smoothed time series coincides with the original time series.

Remark 24.3: Contrary to real time series, for fuzzy regression models

$$\min_{b_1^*, b_2^*, b_3^* \in \mathcal{F}_2(\mathbb{R})} \sum_{t=1}^{5} \rho_2^2 \left(x_t^*, \ b_1^* \oplus b_2^* \cdot t \oplus b_3^* \cdot t^2 \right)$$

with coefficients b_1^*, b_2^* and b_3^* for the generation of a polynomial trend to the first five observations $(x_t^*)_{t=1\,(1)\,5}$, the coefficients $\lambda_{i,t} \in \mathbb{R}$ with

$$b_i^* = \bigoplus_{t=1}^{5} \lambda_{i,t} \cdot x_t^*, \qquad i = 1\,(1)\,3$$

cannot be calculated linearly from $(x_t^*)_{t=1\,(1)\,5}$. A quadratic model yields the above filter $L = (a_{-2}, a_{-1}, a_0, a_1, a_2) = 35^{-1}(-3, 12, 17, 12, -3)$.

The nonexistence of linear estimators in the fuzzy regression model is explained in Körner (1997a, page 69).

This result shows a dilemma: On one hand it would be good to allow negative coefficients in order to have a generalization of the classical theory. On the other hand, the fuzziness of the smoothed values would become too large in the case of allowing negative coefficients.

A solution to this problem which provides a generalization of standard time series, and does not increase the fuzziness in smoothing, is the definition of so-called *extended filters*. Such filters are filtering location and fuzziness separately. By Remark 23.2 the Steiner point of a fuzzy number can be seen as a kind of mean value which is related to location. The observation x_t^* can be described by $x_{t,0}^* = x_t^* \ominus \sigma_{x_t^*}$ where $\sigma_{x_t^*}$ is for the location, and $x_{t,0}^*$ is describing the fuzziness.

Definition 24.2: An *extended linear filter* $L_e = \left((a_{-q}, b_{-q}), \ldots, (a_s, b_s) \right)$, with $(a_i, b_i) \in \mathbb{R} \times \mathbb{R}_0^+, i = -q\,(1)\,s$, is a transformation of the fuzzy time series $(x_t^*)_{t \in T}$

to a series $(y_t^*)_{t=q+1\,(1)\,N-s}$ by

$$y_t^* = L_e \, x_t^* = \left(\sum_{i=-q}^{s} a_i \, \sigma_{x_{t+i}^*} \right) \oplus \left(\bigoplus_{i=-q}^{s} b_i \cdot x_{t+i,0}^* \right), \qquad t = q + 1\,(1)\,N - s \,,$$

where $x_{t,0}^* = x_t^* \ominus \sigma_{x_t^*}$ is the centered fuzzy observation.

24.2.2 Nonlinear filters

If negative coefficients are considered as subtraction of the corresponding observations, another kind of filter can be defined. The basis for this is the so-called *Hukuhara addition*. This addition of fuzzy numbers $x^*, y^* \in \mathcal{F}_2\,(\mathbb{R})$ and $\alpha \in \mathbb{R}$ is defined by

$$x^* \oplus_H \alpha \cdot y^* = \begin{cases} x^* \oplus \alpha \cdot y^* & \text{for} \quad \alpha \geq 0 \\ x^* \sim_H |\alpha| \cdot y^* & \text{for} \quad \alpha < 0 \end{cases}.$$

In order to apply this Hukuhara addition the rules must be declared because this operation is not associative, i.e. for three fuzzy numbers $x^*, y^*, z^* \in \mathcal{F}_2\,(\mathbb{R})$ and real numbers $\alpha, \beta \in \mathbb{R}$ in general $(x^* \oplus_H \alpha \cdot y^*) \oplus_H \beta \cdot z^* \neq x^* \oplus_H (\alpha \cdot y^* \oplus_H \beta \cdot z^*)$. This can be seen in a simple case of interval fuzzy numbers x^* and y^* with characterizing functions $\xi_{x^*}(x) = I_{[-3,3]}(x)$ and $\xi_{y^*}(x) = I_{[-1,1]}(x)$:

$$\underbrace{(x^* \oplus_H (-1) \cdot y^*)}_{= \, 2 \cdot y^*} \oplus_H (-1) \cdot y^* = y^* \neq x^* \oplus_H \underbrace{((-1) \cdot y^* \oplus_H (-1) \cdot y^*)}_{= \, 0} = x^*$$

Moreover the meaning of $\alpha \cdot x^* \oplus_H \beta \cdot y^*$ with $\alpha, \beta \in \mathbb{R}$ has to be defined.

The Hukuhara sum operation $\bigoplus_H$ is defined in the following way: For fuzzy numbers $x_i^* \in \mathcal{F}_2\,(\mathbb{R})$, $i = 1\,(1)\,N$, and real numbers $a_i \in \mathbb{R}$, $i = 1\,(1)\,N$, $a_i^+ = \max\{0, a_i\}$ and $a_i^- = \max\{0, -a_i\}$

$$\bigoplus_{i=1}^{N}{}_H \, a_i \cdot x_i^* = a_1 \cdot x_1^* \oplus_H \ldots \oplus_H a_N \cdot x_N^*$$

$$:= \begin{cases} \left(\displaystyle\bigoplus_{i=1}^{N} a_i^+ \cdot x_i^* \right) \sim_H \left(\displaystyle\bigoplus_{i=1}^{N} a_i^- \cdot x_i^* \right) & \text{for} \quad \displaystyle\sum_{i=1}^{N} a_i^+ \geq \sum_{i=1}^{N} a_i^- \\[4mm] (-1) \cdot \left[\left(\displaystyle\bigoplus_{i=1}^{N} a_i^- \cdot x_i^* \right) \sim_H \left(\displaystyle\bigoplus_{i=1}^{N} a_i^+ \cdot x_i^* \right) \right] & \text{for} \quad \displaystyle\sum_{i=1}^{N} a_i^+ < \sum_{i=1}^{N} a_i^- \end{cases}.$$

First the sums of all fuzzy values with positive and, respectively, negative coefficients are calculated separately. Then the Hukuhara difference of both these sums is calculated, whereby the subtraction depends on the sum of positive and negative coefficients.

Definition 24.3: A transformation $L_H = (a_{-q}, \ldots, a_s) \in \mathbb{R}^{s+q+1}$ of a time series $(x_t^*)_{t \in T}$ to a time series $(y_t^*)_{t=q+1\,(1)\,N-s}$ by

$$y_t^* = L_H x_t^* = \bigoplus_{i=-q}^{s} {}_H \, a_i \cdot x_{t+i}^*, \qquad t = q + 1\,(1)\,N - s,$$

is called a *general filter*.

Lemma 24.1: A general filter is not linear, i.e.

$$L_H(\alpha \cdot x_t^* \oplus \beta \cdot y_t^*) \neq \alpha \cdot L_H x_t^* \oplus \beta \cdot L_H y_t^*.$$

Proof: For fuzzy numbers $x_1^*, x_2^*, y_1^*, y_2^* \in \mathcal{F}_2(\mathbb{R})$ and positive real numbers $\alpha, \beta \in \mathbb{R}^+$ we have in general

$$(\alpha \cdot x_1^* \oplus \alpha \cdot y_1^*) \sim_H (\beta \cdot x_2^* \oplus \beta \cdot y_2^*) \neq (\alpha \cdot x_1^* \sim_H \beta \cdot x_2^*) \oplus (\alpha \cdot y_1^* \sim_H \beta \cdot y_2^*).$$

This can be seen easily for fuzzy numbers of interval type with characterizing functions $\xi_{x_1^*}(x) = I_{[0,1]}(x)$, $\xi_{x_2^*}(x) = I_{[0,3]}(x)$, $\xi_{y_1^*}(x) = I_{[0,2]}(x)$ and $\xi_{y_2^*}(x) = I_{[0,3]}(x)$. For these we have $(2 \cdot x_1^* \oplus 2 \cdot y_1^*) \sim_H (x_2^* \oplus y_2^*) = 0 \neq (2 \cdot x_1^* \sim_H x_2^*) \oplus (2 \cdot y_1^* \sim_H y_2^*) \,\widehat{=}\, I_{[-0.5,0.5]}(\cdot)$. This calculation corresponds to the general filter $L_H = (a_0, a_1) = (2, -1)$ from which for the above observations and $\alpha = \beta = 1$ the result is seen. □

For general filters it is possible to obtain polynomial type time series by filtration. The coefficients of these general filters coincide with the corresponding coefficients of the classical filters in standard time series analysis. For quadratic progress this is explained in the following example.

Example 24.2 As mentioned in Section 24.2.1 for classical real valued time series and filter $L = (a_{-2}, a_{-1}, a_0, a_1, a_2) = 35^{-1}(-3, 12, 17, 12, -3)$ a quadratic progress is unchanged. A general filter L_H with the classical determined coefficients, i.e. $L_H = L$, leaves fuzzy observations with quadratic progress unchanged. This is seen for a time series $(x_t^*)_{t \in T}$ with $x_t^* = b_1^* \oplus b_2^* \cdot t \oplus b_3^* \cdot t^2, t \in T$ and $b_1^*, b_2^*, b_3^* \in \mathcal{F}_2(\mathbb{R})$ in the following way:

$$y_t^* = L_H x_t^* = \bigoplus_{i=-2}^{2} {}_H \, a_i \cdot x_{t+i}^*$$

$$= \left(\frac{12}{35} \cdot x_{t-1}^* \oplus \frac{17}{35} \cdot x_t^* \oplus \frac{12}{35} \cdot x_{t+1}^* \right) \sim_H \left(\frac{3}{35} \cdot x_{t-2}^* \oplus \frac{3}{35} \cdot x_{t+2}^* \right)$$

$$= \left(\frac{41}{35} \cdot b_1^* \oplus \frac{41\,t}{35} \cdot b_2^* \oplus \frac{41\,t^2 + 24}{35} \cdot b_3^* \right) \sim_H \left(\frac{6}{35} \cdot b_1^* \oplus \frac{6\,t}{35} \cdot b_2^* \oplus \frac{6\,t^2 + 24}{35} \cdot b_3^* \right)$$

By

$$\left(\frac{6}{35} \cdot b_1^* \oplus \frac{6\,t}{35} \cdot b_2^* \oplus \frac{6\,t^2 + 24}{35} \cdot b_3^*\right) \oplus \left(b_1^* \oplus b_2^* \cdot t \oplus b_3^* \cdot t^2\right)$$

$$= \frac{41}{35} \cdot b_1^* \oplus \frac{41\,t}{35} \cdot b_2^* \oplus \frac{41\,t^2 + 24}{35} \cdot b_3^*$$

we have $y_t^* = b_1^* \oplus b_2^* \cdot t \oplus b_3^* \cdot t^2$, and by Theorem 23.1 this is uniquely determined. The smoothed time series therefore has quadratic progress.

24.3   Exponential smoothing

The recursively defined filters of exponential smoothing provides the possibility of inclusion of additional observations in a simple way. By using the minimum t-norm it is possible to determine the arithmetic mean $\bar{x}_N^*$ of observations $(x_t^*)_{t \in T}$ recursively. If the mean value $\bar{x}_N^*$ is given and an additional observation x_{N+1}^* is observed, the mean value $\bar{x}_{N+1}^*$ can be calculated as

$$\bar{x}_{N+1}^* = \frac{1}{N+1} \cdot \bigoplus_{t=1}^{N+1} x_t^* = \frac{N}{N+1} \cdot \bar{x}_N^* \oplus \left(1 - \frac{N}{N+1}\right) \cdot x_{N+1}^*.$$

If $\frac{N}{N+1}$ is replaced by a general weight β with $0 < \beta < 1$ a *filter of simple exponential smoothing* is obtained.

In exponential smoothing the values y_t^*, $t = 2\,(1)\,N$, of the transformed series $(y_t^*)_{t \in T}$ of fuzzy data $(x_t^*)_{t \in T}$ as above are calculated recursively from x_t^* and the smoothed value y_{t-1}^* by

$$y_t^* = \beta \cdot y_{t-1}^* \oplus (1 - \beta) \cdot x_t^* = (1 - \beta) \cdot \bigoplus_{i=0}^{t-2} \beta^i \cdot x_{t-i}^* \oplus \beta^{t-1} \cdot x_1^*, \quad t = 2\,(1)\,N,$$

with $y_1^* = x_1^*$. The smoothing parameter β is controlling the smoothing behavior. For small β the last observation x_t^* has more influence, whereas β near 1 produces more smooth transformed series $(y_t^*)_{t \in T}$.

Similar to classical time series an ideal smoothing parameter can be determined using a one-step prediction $\widehat{x}_{N,1}^*(\beta)$. This one-step prediction for x_{N+1}^* based on the observed values $x_1^*, \ldots, x_N^*$ and parameter β is, analogous to the classical case, obtained from the foregoing estimate $\widehat{x}_{N-1,1}^*$ and the last observation x_N^*, recursively by

$$\widehat{x}_{N,1}^*(\beta) = \beta \cdot \widehat{x}_{N-1,1}^*(\beta) \oplus (1 - \beta) \cdot x_N^* = (1 - \beta) \cdot \bigoplus_{i=0}^{N-2} \beta^i \cdot x_{N-i}^* \oplus \beta^{N-1} \cdot \widehat{x}_{1,1}^*(\beta).$$

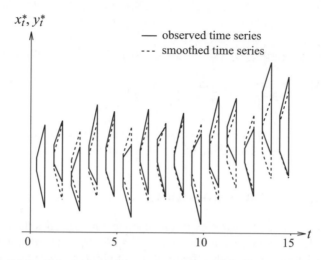

Figure 24.2   Exponential smoothing with optimal smoothing parameter.

In practical applications the following is assumed: $\widehat{x}^*_{1,1}(\beta) = x^*_1$.

In order to find a suitable smoothing parameter β different values are used for one-step predictions and compared with observed values. Then the value $\widehat{\beta}$ is used as a smoothing parameter which minimizes the sum of squared distances between predictions and observations:

$$\sum_{t=1}^{N-1} \rho_2^2 \left(\widehat{x}^*_{t,1}(\widehat{\beta}), x^*_{t+1}\right) = \min_{\beta} \sum_{t=1}^{N-1} \rho_2^2 \left(\widehat{x}^*_{t,1}(\beta), x^*_{t+1}\right).$$

Example 24.3 In Figure 24.2 the exponential smoothing of the data from Example 24.1 with the method described above using $\widehat{\beta}$ is depicted.

24.4   Components model

For this section it is assumed that all data of the time series $(x^*_t)_{t \in T}$ have characterizing function with compact support. For practical applications this is usually fulfilled. Therefore this is no restriction.

In classical descriptive time series analysis the observed data $(x_t)_{t \in T}$ are assumed to be generated by the following model:

$$x_t = m_t + s_t + \varepsilon_t, \qquad t = 1\,(1)\,N. \tag{24.2}$$

The component m_t is called a *trend* and describes the long term behavior of the time series. The component s_t is called a *seasonal component*, and describes the cyclic behavior with more or less constant time period. The last component ε_t, the so-called *error term*, is modelling the stochastic influences. The time period of the seasonal component is denoted by p. For so-called *stable seasonal behavior* it is assumed that $s_t = s_{t+kp}, k \in \mathbb{Z}$.

In the case of fuzzy data $(x_t^*)_{t \in T}$ the components of model (24.2) are also fuzzy. The model takes the following form:

$$x_t^* \approx m_t^* \oplus s_t^*, \qquad t = 1\,(1)\,N.$$

For the seasonal component s_t^* it is also assumed that $s_t^* = s_{t+kp}^*, k \in \mathbb{Z}$. Contrary to standard time series analysis, for fuzzy data x_t^* no exact decomposition into trend, seasonal component, and error term is possible. The reason for that is the semilinearity of $\mathcal{F}(\mathbb{R})$.

Therefore for some times t no fuzzy number ε_t^* can be found such that $x_t^* = m_t^* \oplus s_t^* \oplus \varepsilon_t^*$ for given fuzzy values m_t^* and s_t^*. This follows for t with $supp\,(x_t^*) \subseteq supp\,(m_t^* \oplus s_t^*)$. For this reason the model is approximative.

An example is the following. In a region of winter tourism the total gain from tourism can be estimated only as fuzzy. In summer, with less guests, the income is much smaller and therefore the estimate can be more precise.

Contrary to moving averages and exponential smoothing for the estimation of fuzzy trends and fuzzy seasonal components the application of the extension principle is not suitable. The reason for this is that by application of the extension principle no real decomposition is made. Moreover, in particular, the sum $m_t^* \oplus s_t^*$ has too large fuzziness. Therefore an alternative solution is necessary.

24.4.1 Model without seasonal component

If no seasonal influence exists it can be assumed that $s_t^* \equiv 0, t \in T$, and the determination of m_t^* is, similar to the classical situation, a fuzzy regression problem for the values $(x_t^*)_{t \in T}$. The estimation $\widehat{x}_{N_f}^*$ for $x_{N_f}^*$ at time $N_f > N$ is given by $\widehat{x}_{N_f}^* = \widehat{m}_{N_f}^*$, i.e. the value of the regression trend.

24.4.2 Model with seasonal component

For this kind of model the calculation of the components m_t^* or s_t^* is far more complicated than in the classical case. The problems come from the semilinear structure of $\mathcal{F}(\mathbb{R})$. In the classical situation moving averages over a period are used to remove seasonal effects. As estimate of the seasonal component for a time t the difference between the observed value x_t and the moving average is used.

In the case of fuzzy data this method is not suitable. First, the fuzziness of the individual values is increasing for sums. Secondly, an estimation by the Hukuhara difference between observed value x_t^* and the moving average is problematic because for some time points the fuzziness of the observation is smaller than the fuzziness of the mean value. Using the Hukuhara difference would yield a precise number as an estimation.

A method which reduces the determination of the fuzzy components to the methods of classical time series is the following: The seasonal influence on the fuzziness of the observations $(x_t^*)_{t \in T}$ is recognized in the length of the δ-cuts $C_\delta(x_t^*) = [\underline{x}_{t,\delta}, \overline{x}_{t,\delta}]$. The length of the δ-cuts

$$l_{t,\delta} = l_{C_\delta(x_t^*)} = \overline{x}_{t,\delta} - \underline{x}_{t,\delta} = \widetilde{m}_{t,\delta} + \widetilde{s}_{t,\delta} + \widetilde{\varepsilon}_{t,\delta} \in \mathbb{R}_0^+$$

of the elements $(x_t^*)_{t \in T}$ for every $\delta \in (0; 1]$ can be considered as standard time series $(l_{t,\delta})_{t \in T}$ with seasonal component. Conditions for the trend component $\widetilde{m}_{t,\delta}$ and the seasonal component $\widetilde{s}_{t,\delta}$ are $\widetilde{m}_{t,\delta} \geq 0$ and $\widetilde{s}_{t,\delta} \geq 0$, respectively. These conditions can be fulfilled because the following holds: $l_{t,\delta} \in \mathbb{R}_0^+$. In order to provide the uniqueness of the components $\widetilde{m}_{t,\delta}$ and $\widetilde{s}_{t,\delta}$ different conditions can be used.

In the following, the condition

$$\min_{t=1(1)N} \widetilde{m}_{t,\delta} = 0 \tag{24.3}$$

is used. Other possibilies are $\min_{t=1(1)N} \widetilde{s}_{t,\delta} = 0$ or a mixture of similar conditions. By using (24.3) a larger part of the fuzziness of the observations is ascribed to the seasonal component.

The quantities $\widetilde{m}_{t,\delta}$ and $\widetilde{s}_{t,\delta}$ are influenced by the fuzziness of the individual observations and not by their position. Therefore they can be used to estimate the fuzziness of trend and seasonal components. By Remark 23.3 the Steiner point σ_{x^*} of a fuzzy number x^* is a kind of midpoint. Therefore σ_{x^*} can be used as information on the position, and $x_0^* = x^* \ominus \sigma_{x^*}$ can be used as information on the fuzziness of x^*.

In order to determine the trend and seasonal component first for all values of the fuzzy time series $(x_t^*)_{t \in T}$ the Steiner point $\sigma_{x_t^*}$ is calculated. From the classical time series of the Steiner points $(\sigma_{x_t^*})_{t \in T}$ with standard time series methods the position m_t and s_t of the trend and seasonal component can be calculated. The calculation of the fuzziness $m_{t,0}^*$ and $s_{t,0}^*$ of the components is done from the time series $(x_{t,0}^*)_{t \in T} = (x_t^* \ominus \sigma_{x_t^*})_{t \in T}$. With this decomposition the whole component model reads

$$x_t^* \approx (m_t + s_t + \varepsilon_t) \oplus (m_{t,0}^* \oplus s_{t,0}^*), \qquad t = 1\,(1)\,N. \tag{24.4}$$

The decomposition into position and fuzziness does not change the length of the δ-cuts, i.e. $l_{C_\delta(x_t^*)} = l_{C_\delta(x_{t,0}^*)} = l_{t,\delta}$. The components $\widetilde{m}_{t,\delta}$, $\widetilde{s}_{t,\delta}$, and $\widetilde{\varepsilon}_{t,\delta}$ of the classical time series $(l_{t,\delta})_{t \in T}$ from the beginning of this section and the δ-cuts $C_\delta\left(\widehat{m}_{t,0}^*\right)$ and $C_\delta\left(\widehat{s}_{t,0}^*\right)$ are used to estimate $m_{t,0}^*$ and $s_{t,0}^*$ in the following way:

$$C_\delta\left(\widehat{m}_{t,0}^*\right) = \frac{\widetilde{m}_{t,\delta}}{l_{t,\delta}} C_\delta(x_{t,0}^*) \qquad \text{and} \qquad C_\delta\left(\widehat{s}_{t,0}^*\right) = \frac{\widetilde{s}_{t,\delta}}{l_{t,\delta}} C_\delta(x_{t,0}^*) \quad \text{if } l_{t,\delta} \neq 0$$

with $C_\delta\left(\widehat{m}_{t,0}^*\right) = C_\delta\left(\widehat{s}_{t,0}^*\right) = [0;0] = 0$ if $l_{t,\delta} = 0$.

The above decomposition of the δ-cuts follows the consideration after which the value $\widetilde{m}_{t,\delta}$ describes the influence of the trend, and $\widetilde{s}_{t,\delta}$ the influence of the seasonal component on the length of the δ-cut $C_\delta(x_t^*)$.

This method has the following problems: Without additional conditions concerning the quantities $\widetilde{m}_{t,\delta}$ and $\widetilde{s}_{t,\delta}$ the δ-cuts of the estimations for the components can be such that $C_{\delta_2}\left(\widehat{m}_{t,0}^*\right) \not\subseteq C_{\delta_1}\left(\widehat{m}_{t,0}^*\right)$ and $C_{\delta_2}\left(\widehat{s}_{t,0}^*\right) \not\subseteq C_{\delta_1}\left(\widehat{s}_{t,0}^*\right)$ for $0 < \delta_1 < \delta_2 \leq 1$. Additional conditions in order to avoid this are in general complicated and increase the computational work for the determination of the components.

A similar but simpler method is the following: For the determination of the fuzziness of the trend and seasonal component not all δ-cuts are considered but the 'average' length of fuzziness. Instead of the time series $(l_{t,\delta})_{t\in T}$, $\delta \in (0; 1]$, the time series

$$(l_t)_{t\in T} = \left(\int_0^1 l_{t,\delta}\, d\delta\right)_{t\in T}$$

is considered. $(l_t)_{t\in T}$ is a time series with seasonal influence, i.e.

$$l_t = \widetilde{m}_t + \widetilde{s}_t + \widetilde{\varepsilon}_t \in \mathbb{R}_0^+, \qquad t = 1\,(1)\,N,$$

with $\widetilde{m}_t \geq 0$ and $\widetilde{s}_t \geq 0$. From these two values estimates of the fuzziness of the trend and seasonal component are obtained by

$$\widehat{m}_{t,0}^* = \frac{\widetilde{m}_t}{l_t}\cdot x_{t,0}^* \qquad \text{and} \qquad \widehat{s}_{t,0}^* = \frac{\widetilde{s}_t}{l_t}\cdot x_{t,0}^* \qquad \text{if } l_t \neq 0$$

for $t \in T$. For $l_t = 0$ the estimates are $\widehat{m}_{t,0}^* = \widehat{s}_{t,0}^* = 0$. Unfortunately for these estimates the condition $\widehat{s}_{t,0}^* = \widehat{s}_{t+kp,0}^*$, $k \in \mathbb{Z}$, is not necessarily fulfilled.

The final estimate of the seasonal component $s_{t,0}^*$ is the average of all elements $\widehat{s}_{t+kp,0}^*$, $k \in \mathbb{Z}$. The fuzzy trend $m_{t,0}^*$ is calculated using a regression of the quantities $\widehat{m}_{t,0}^*$. The estimate $\widehat{x}_{N_f}^*$ for $x_{N_f}^*$ for time $N_f > N$ is

$$\widehat{x}_{N_f}^* = (\widehat{m}_{N_f} + \widehat{s}_{N_f}) \oplus \widehat{m}_{N_f}^* \oplus \widehat{s}_{N_f}^* = (\widehat{m}_{N_f} \oplus \widehat{m}_{N_f}^*) \oplus (\widehat{s}_{N_f} \oplus \widehat{s}_{N_f}^*),$$

where $\widehat{m}_{N_f}$ and $\widehat{s}_{N_f}$ are the estimates calculated from the real valued time series $\left(\sigma_{x_t^*}\right)_{t\in T}$, and $\widehat{m}_{N_f}^*$ and $\widehat{s}_{N_f}^*$ are the estimates of the fuzziness. The quantity $(\widehat{m}_{N_f} \oplus \widehat{m}_{N_f}^*)$ is the fuzzy trend and $(\widehat{s}_{N_f} \oplus \widehat{s}_{N_f}^*)$ is the fuzzy seasonal component.

Example 24.4 Table 24.1 contains 48 trapezoidal fuzzy observations with seasonal component.

The values are depicted in Figure 24.3.

First, for every time the Steiner point $\sigma_{x_t^*}$ and the average length l_t of the δ-cuts of the observation are calculated. In Figure 24.4 the time series $\left(\sigma_{x_t^*}\right)_{t\in T}$ of the Steiner points is depicted as a solid line, and the time series $(l_t)_{t\in T}$ of average lengths of the δ-cuts is depicted as a dotted line.

Table 24.1   Values of 48 trapezoidal fuzzy observations with seasonal component.

$x_1^* = t^*(1.35, 0.20, 0.44, 0.56)$ $x_2^* = t^*(4.84, 0.16, 0.62, 0.74)$
$x_3^* = t^*(7.94, 0.26, 0.67, 1.25)$ $x_4^* = t^*(9.57, 0.23, 0.62, 1.24)$
$x_5^* = t^*(7.41, 0.28, 0.74, 0.85)$ $x_6^* = t^*(4.37, 0.25, 0.84, 0.84)$
$x_7^* = t^*(0.62, 0.16, 0.59, 0.66)$ $x_8^* = t^*(-1.49, 0.16, 0.70, 0.48)$
$x_9^* = t^*(-1.47, 0.10, 0.56, 0.44)$ $x_{10}^* = t^*(-1.41, 0.11, 0.29, 0.43)$
$x_{11}^* = t^*(-1.42, 0.12, 0.34, 0.57)$ $x_{12}^* = t^*(-0.67, 0.13, 0.46, 0.58)$
$x_{13}^* = t^*(1.72, 0.19, 0.97, 0.63)$ $x_{14}^* = t^*(5.45, 0.18, 0.55, 1.20)$
$x_{15}^* = t^*(9.74, 0.27, 1.28, 0.98)$ $x_{16}^* = t^*(11.88, 0.26, 0.72, 0.77)$
$x_{17}^* = t^*(9.82, 0.27, 0.66, 1.14)$ $x_{18}^* = t^*(6.46, 0.26, 0.85, 1.29)$
$x_{19}^* = t^*(2.65, 0.20, 0.81, 0.74)$ $x_{20}^* = t^*(0.68, 0.18, 0.79, 0.49)$
$x_{21}^* = t^*(0.08, 0.12, 0.46, 0.46)$ $x_{22}^* = t^*(0.49, 0.10, 0.49, 0.35)$
$x_{23}^* = t^*(0.33, 0.15, 0.58, 0.61)$ $x_{24}^* = t^*(1.21, 0.14, 0.38, 0.74)$
$x_{25}^* = t^*(3.64, 0.22, 0.98, 0.52)$ $x_{26}^* = t^*(7.24, 0.28, 1.17, 1.20)$
$x_{27}^* = t^*(10.67, 0.34, 0.80, 0.81)$ $x_{28}^* = t^*(11.88, 0.28, 0.92, 0.98)$
$x_{29}^* = t^*(10.99, 0.30, 1.21, 1.36)$ $x_{30}^* = t^*(8.20, 0.24, 0.72, 0.74)$
$x_{31}^* = t^*(4.66, 0.23, 0.67, 0.86)$ $x_{32}^* = t^*(2.23, 0.19, 0.82, 0.84)$
$x_{33}^* = t^*(1.81, 0.16, 0.64, 0.51)$ $x_{34}^* = t^*(2.46, 0.16, 0.53, 0.35)$
$x_{35}^* = t^*(2.32, 0.18, 0.60, 0.40)$ $x_{36}^* = t^*(2.51, 0.19, 0.62, 0.56)$
$x_{37}^* = t^*(6.04, 0.22, 0.74, 0.73)$ $x_{38}^* = t^*(10.34, 0.26, 1.15, 0.96)$
$x_{39}^* = t^*(12.45, 0.35, 0.93, 1.43)$ $x_{40}^* = t^*(14.80, 0.27, 0.75, 1.36)$
$x_{41}^* = t^*(13.17, 0.33, 1.38, 1.34)$ $x_{42}^* = t^*(8.86, 0.26, 0.96, 1.16)$
$x_{43}^* = t^*(6.02, 0.26, 1.10, 0.86)$ $x_{44}^* = t^*(4.22, 0.18, 0.69, 0.74)$
$x_{45}^* = t^*(3.46, 0.19, 0.53, 0.47)$ $x_{46}^* = t^*(3.58, 0.19, 0.55, 0.47)$
$x_{47}^* = t^*(3.92, 0.16, 0.67, 0.46)$ $x_{48}^* = t^*(4.76, 0.21, 0.50, 0.77)$

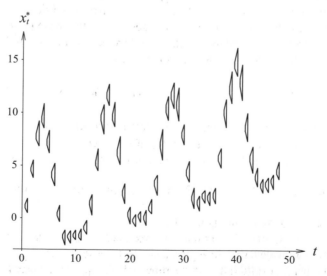

Figure 24.3   Fuzzy observations with seasonal influence.

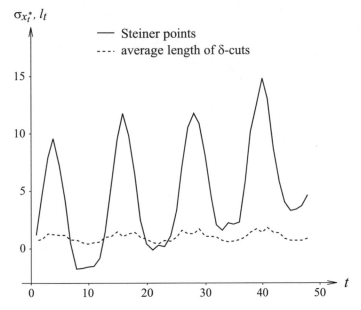

Figure 24.4   Time series of Steiner points and average length of δ-cuts.

From Figure 24.4 it can be seen, that the length of the period of both time series is 12. For both time series with classical methods the trend and seasonal component are calculated. After elimination of the seasonal component, both $(\sigma_{x_t^*})_{t=1(1)48}$ and $(l_t)_{t=1(1)48}$ show approximately linear behavior. To these values using regression models a line is attached.

From this the fuzzy seasonal components $s_t^*, t = 1\,(1)\,48$, with $s_t^* = s_{t+12k}^*, k \in \mathbb{Z}$, are obtained. These are given in Table 24.2 and depicted in Figure 24.5.

The fuzzy linear trend is

$$m_t^* = t^*(1.59, 0.043, 0.16, 0.19) + t^*(0.14, 0.0018, 0.0061, 0.0095) \cdot t \,.$$

In Figure 24.6 the observed data $x_t^*, t = 1\,(1)\,48$, and the predicted values $\widehat{x}_t^* = m_t^* \oplus s_t^*, t = 1\,(1)\,48$, are depicted for comparison.

Table 24.2   Values of the fuzzy seasonal components.

$s_1^* = t^*(-1.30, 0.13, 0.46, 0.37)$	$s_7^* = t^*(-1.55, 0.12, 0.44, 0.44)$
$s_2^* = t^*(2.47, 0.16, 0.61, 0.75)$	$s_8^* = t^*(-3.90, 0.10, 0.42, 0.35)$
$s_3^* = t^*(5.58, 0.22, 0.65, 0.78)$	$s_9^* = t^*(-4.36, 0.05, 0.20, 0.18)$
$s_4^* = t^*(7.36, 0.17, 0.48, 0.69)$	$s_{10}^* = t^*(-4.12, 0.03, 0.11, 0.10)$
$s_5^* = t^*(5.70, 0.22, 0.72, 0.87)$	$s_{11}^* = t^*(-4.36, 0.06, 0.21, 0.20)$
$s_6^* = t^*(2.09, 0.16, 0.54, 0.63)$	$s_{12}^* = t^*(-3.88, 0.07, 0.19, 0.26)$

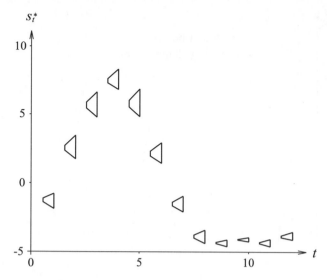

Figure 24.5   Fuzzy seasonal components.

24.5   Difference filters

In matching polynomials to time series $(x_t^*)_{t \in T}$ a problem is the degree of the polynomial. Similar to classical time series it is also possible for fuzzy time series using so-called difference filters.

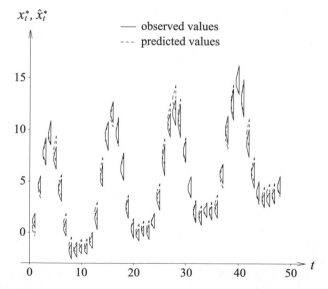

Figure 24.6   Observed and predicted values of the time series.

Remark 24.4: For a polynomial $f(t) = a_0^* \oplus a_1^* \cdot t \oplus a_2^* \cdot t^2 \oplus \ldots \oplus a_m^* \cdot t^m$ with degree $m > 0$ and $a_0^*, a_1^*, \ldots, a_m^* \in \mathcal{F}_2(\mathbb{R})$ the difference $g(t) = f(t) \sim_H f(t-1)$ is for $t > 0$ a polynomial of maximal degree $m-1$.

Proof: By Remark 23.6 with

$$g(t) = a_1^* \oplus a_2^* \cdot \left(t^2 - (t-1)^2 \right) \oplus \ldots \oplus a_m^* \cdot \left(t^m - (t-1)^m \right)$$

and $t^i - (t-1)^i \geq 0, i = 1\,(1)\,m$ we obtain

$$g(t) \oplus f(t-1) = a_1^* \oplus a_2^* \cdot \left(t^2 - (t-1)^2 \right) \oplus \ldots \oplus a_m^* \cdot \left(t^m - (t-1)^m \right) \oplus$$
$$a_0^* \oplus a_1^* \cdot (t-1) \oplus a_2^* \cdot (t-1)^2 \oplus \ldots \oplus a_m^* \cdot (t-1)^m = f(t).$$

The Hukuhara difference $g(t)$ is a polynomial of degree $(m-1)$ which is uniquely determined by Theorem 23.1. $\qquad\square$

Definition 24.4: Let $(x_t^*)_{t \in T}$ be a fuzzy time series. A mapping Δ, defined by

$$\Delta x_t^* = x_t^* \sim_H x_{t-1}^*, \qquad t = 2\,(1)\,N,$$

is called a *difference filter of order* 1. Difference filters of order m with $m > 1$ are recursively defined by

$$\Delta^m x_t^* = \left(\Delta^{m-1} x_t^* \right) \sim_H \left(\Delta^{m-1} x_{t-1}^* \right), \qquad t = m+1\,(1)\,N.$$

The difference operation reduces the degree of a polynomial by 1. Applying the difference operation $(m-1)$ times to a polynomial of degree m results in constant values.

Remark 24.5: A difference filter in the context of Definition 24.3 is a general filter $L_H(a_{-1}, a_0) = (-1, 1)$.

Similar to standard time series difference filters can be used to eliminate seasonal effects. In order to do this so-called *seasonal differences of length p*, defined by

$$\Delta_p x_t^* = x_t^* \sim_H x_{t-p}^*, \qquad t = p+1\,(1)\,N,$$

are used.

It is possible to combine the difference operation with seasonal differences. If a fuzzy time series $(x_t^*)_{t \in T}$ has seasonal oscillations with period p and linear trend, than the filtered series $y_t^* = \Delta \left(\Delta_p x_t^* \right)$, where first a seasonal filter of length p is applied, and then a difference filter, is approximately constant.

For numerical calculation of the length of the period and the trend, for different natural numbers p and m the hypothesis

$$\Delta^m \left(\Delta_p x_t^* \right) \equiv \text{const}$$

can be tested.

Some references for testing hypotheses based on fuzzy data are Arnold (1996), Grzegorzewski (2001), Körner (2000) and Römer and Kandel (1995).

24.6   Generalized Holt–Winter method

For details of the Holt–Winter method in the case of standard time series see Janacek (2001).

The Holt–Winter method for standard time series $(x_t)_{t \in T}$ assumes a component model

$$x_t = m_t + s_t + \varepsilon_t, \qquad t = 1\,(1)\,N,$$

with local linear trend $m_t = a + b\,t$ and period length p of the seasonal component s_t. If for time $N - 1$ estimators $\widehat{m}_{N-1}, \widehat{b}_{N-1}$ and $\widehat{s}_{N-p}$ are given, then the estimates for time N can be calculated recursively:

$$\widehat{b}_N = (1 - \alpha)\,(\widehat{m}_{N-1} - \widehat{m}_{N-2}) + \alpha\,\widehat{b}_{N-1} \qquad (24.5)$$

$$\widehat{m}_N = (1 - \beta)\,\left(x_N - \widehat{s}_{N-p}\right) + \beta\,\left(\widehat{m}_{N-1} + \widehat{b}_N\right) \qquad (24.6)$$

$$\widehat{s}_N = (1 - \gamma)\,(x_N - \widehat{m}_N) + \gamma\,\widehat{s}_{N-p} \qquad (24.7)$$

Starting values for the recursion from the first p-values of the time series are calculated as follows:

$$\widehat{m}_p = \frac{1}{p} \sum_{t=1}^{p} x_t, \quad \widehat{b}_p = 0 \quad \text{and} \quad \widehat{s}_t = x_t - \widehat{m}_p, \; t = 1\,(1)\,p \qquad (24.8)$$

The value $\widehat{b}_p = 0$ follows the assumption that at the beginning no trend is present. A h-step prediction at time N, with $h > 0$, is given by

$$\widehat{x}_{N,h}(\alpha, \beta, \gamma) = \begin{cases} \widehat{m}_N + h\,\widehat{b}_N + \widehat{s}_{N+h-p} & h = 1\,(1)\,p \\ \widehat{m}_N + h\,\widehat{b}_N + \widehat{s}_{N+h-2p} & h = p + 1\,(1)\,2\,p \\ \;\;\vdots & \;\;\vdots \end{cases}.$$

There are different approaches for the selection of the smoothing parameters α, β and γ. Most frequently the one-step predictions $\widehat{x}_{t,1}(\alpha, \beta, \gamma)$ are compared with the data $(x_t)_{t \in T}$ and the values $\tilde{\alpha}, \tilde{\beta}$ and $\tilde{\gamma}$ obeying

$$\sum_{t=p}^{N-1} \left(x_{t+1} - \widehat{x}_{t,1}(\tilde{\alpha}, \tilde{\beta}, \tilde{\gamma})\right)^2 = \min_{\alpha, \beta, \gamma} \sum_{t=p}^{N-1} \left(x_{t+1} - \widehat{x}_{t,1}(\alpha, \beta, \gamma)\right)^2$$

are used.

In the case of fuzzy data $(x_t^*)_{t \in T}$ a model of the form

$$x_t^* \approx m_t^* \oplus s_t^*, \qquad t = 1\,(1)\,N,$$

with linear trend $m_t^* = a^* \oplus b^* \cdot t$ is assumed.

For given estimates $\widehat{m}_{N-1}^*, \widehat{b}_{N-1}^*$ and $\widehat{s}_{N-1}^*$ at time $N-1$, estimates $\widehat{m}_N^*, \widehat{b}_N^*$ and $\widehat{s}_N^*$ can be calculated using Equations (24.5), (24.6) and (24.7):

$$\widehat{b}_N^* = (1 - \alpha) \cdot (\widehat{m}_{N-1}^* \sim_H \widehat{m}_{N-2}^*) \oplus \alpha \cdot \widehat{b}_{N-1}^*$$
$$\widehat{m}_N^* = (1 - \beta) \cdot (x_N^* \sim_H \widehat{s}_{N-p}^*) \oplus \beta \cdot (\widehat{m}_{N-1}^* \oplus \widehat{b}_N^*)$$
$$\widehat{s}_N^* = (1 - \gamma) \cdot (x_N^* \sim_H \widehat{m}_N^*) \oplus \gamma \cdot \widehat{s}_{N-p}^*$$

Starting values can be obtained as described in Section 24.4 from the first p-values by decomposition into position and fuzziness component. From (24.3) providing uniqueness of the decomposition follows that the estimate for $\widehat{m}_p^*$ is the average of the Steiner points $(\sigma_{x_t^*})_{t=1(1)p}$, and therefore a real number. The seasonal components are estimated by

$$\widehat{s}_t^* = x_t^* \ominus \widehat{m}_p^*, \qquad t = 1\,(1)\,p.$$

Analogously to the classical case $b_p^* = 0$ is used. A prediction $\widehat{x}_{N+h}^*$ for x_{N+h}^*, $h > 0$, at time N is calculated by

$$\widehat{x}_{N,h}^*(\alpha, \beta, \gamma) = \begin{cases} \widehat{m}_N^* \oplus h \cdot \widehat{b}_N^* \oplus \widehat{s}_{N+h-p}^* & h = 1\,(1)\,p \\ \widehat{m}_N^* \oplus h \cdot \widehat{b}_N^* \oplus \widehat{s}_{N+h-2p}^* & h = p+1\,(1)\,2\,p \\ \vdots & \vdots \end{cases}.$$

The determination of suitable smoothing parameters $\tilde{\alpha}, \tilde{\beta}$ and $\tilde{\gamma}$ is possible as in the classical case by

$$\bigoplus_{t=p}^{N-1} \rho_2^2 \left(x_{t+1}^*, \widehat{x}_{t,1}^*(\tilde{\alpha}, \tilde{\beta}, \tilde{\gamma})\right) = \min_{\alpha, \beta, \gamma} \bigoplus_{t=p}^{N-1} \rho_2^2 \left(x_{t+1}^*, \widehat{x}_{t,1}^*(\alpha, \beta, \gamma)\right). \qquad (24.9)$$

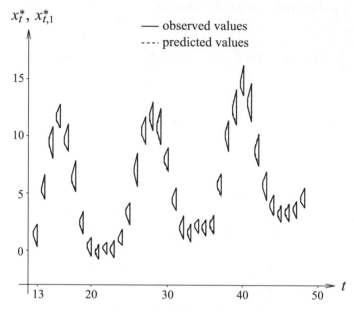

Figure 24.7   Observed and predicted values.

Example 24.5 The generalized Holt–Winter method is applied to the 48 observations from example 24.4. The smoothing parameters from (24.9) are $\alpha = 0.95$, $\beta = 0.31$ and $\gamma = 0.38$. In Figure 24.7 both, the observed data as well as the one-step predictions using optimal smoothing parameters are depicted.

24.7   Presentation in the frequency domain

For an absolute summable sequence $(x_t)_{t \in \mathbb{Z}}$ of real numbers and the functions

$$C(\lambda) = \sum_{t \in \mathbb{Z}} x_t \, \cos(2 \pi \lambda t) \qquad \text{and} \qquad S(\lambda) = \sum_{t \in \mathbb{Z}} x_t \, \sin(2 \pi \lambda t)$$

the inverse of the Fourier transformation exists and the following holds:

$$x_t = 2 \int_0^{0.5} C(\lambda) \, \cos(2 \pi \lambda t) \, d\lambda + 2 \int_0^{0.5} S(\lambda) \, \sin(2 \pi \lambda t) \, d\lambda \,.$$

Therefore the consideration of sequences in the time domain is equivalent to that in the frequency domain. The Fourier transformation contains the same information as the series itself. In the case of finite time series $(x_t)_{t \in T}$ it is sufficient to know the Fourier transformation for finitely many frequency values. A time series with an odd number of observations $N = 2M + 1$ can be represented uniquely using the so-called *Fourier frequencies* $\lambda_k = \frac{k}{N}$, $k = 0\,(1)\,M$:

$$C(\lambda_k) = \frac{1}{N} \sum_{t=1}^{N} x_t \cos(2\pi\,\lambda_k\,t), \qquad k = 0\,(1)\,M,$$

and

$$S(\lambda_k) = \frac{1}{N} \sum_{t=1}^{N} x_t \sin(2\pi\,\lambda_k\,t), \qquad k = 1\,(1)\,M,$$

implies

$$x_t = C(\lambda_0) + 2 \sum_{k=1}^{M} \Big(C(\lambda_k)\,\cos(2\pi\,\lambda_k\,t) + S(\lambda_k)\,\sin(2\pi\,\lambda_k\,t) \Big). \qquad (24.10)$$

For a fuzzy time series $(x_t^*)_{t \in T}$ such a presentation does not exist in general. The reason for that is the semilinearity of the structure $(\mathcal{F}(\mathbb{R}), \oplus, \cdot)$.

Another possibility would be using the Hukuhara addition. Examples show that uniqueness cannot be achieved for fuzzy time series.

Example 24.6 For three observations $x_1^*, x_2^*, x_3^* \in \mathcal{F}_2(\mathbb{R})$ with corresponding characterizing functions $\xi_{x_1^*}(x) = I_{[-1,1]}(x)$, $\xi_{x_2^*}(x) = I_{[-1,1]}(x)$ and $\xi_{x_3^*}(x) = I_{\{0\}}(x)$ as well as frequencies $\lambda_k = \frac{k}{3}$, $k = -1, 0, 1$, there exist no fuzzy coefficients $C^*(\lambda_k)$, $k = -1, 0, 1$, and $S^*(\lambda_k)$, $k = -1, 0, 1$, with

$$x_t^* = \bigoplus_{k=-1}^{1}{}_{H}\ C^*(\lambda_k) \cdot \cos(2\pi\,\lambda_k\,t)\ \oplus_H \bigoplus_{k=-1}^{1}{}_{H}\ S^*(\lambda_k) \cdot \sin(2\pi\,\lambda_k\,t), \qquad t = 1\,(1)\,3.$$

Proof: For the time points $t = 1, 2, 3$ the following equations hold:

$t = 1:$
$$\begin{aligned}
x_1^* &= \left(-\tfrac{1}{2}\right) \cdot C^*(\lambda_{-1})\ \oplus_H\ C^*(\lambda_0)\ \oplus_H\ \left(-\tfrac{1}{2}\right) \cdot C^*(\lambda_1)\ \oplus_H \\
&\quad \left(-\tfrac{\sqrt{3}}{2}\right) \cdot S^*(\lambda_{-1})\ \oplus_H\ \tfrac{\sqrt{3}}{2} \cdot S^*(\lambda_1) \\
&= \left(C^*(\lambda_0) \oplus \tfrac{\sqrt{3}}{2} \cdot S^*(\lambda_1) \right) \sim_H \left(\tfrac{1}{2} \cdot C^*(\lambda_{-1}) \oplus \tfrac{1}{2} \cdot C^*(\lambda_1) \oplus \tfrac{\sqrt{3}}{2} \cdot S^*(\lambda_{-1}) \right)
\end{aligned}$$

$t = 2$:

$$
\begin{aligned}
x_2^* &= \left(-\tfrac{1}{2}\right) \cdot C^*(\lambda_{-1}) \oplus_H C^*(\lambda_0) \oplus_H \left(-\tfrac{1}{2}\right) \cdot C^*(\lambda_1) \oplus_H \\
&\quad \tfrac{\sqrt{3}}{2} \cdot S^*(\lambda_{-1}) \oplus_H \left(-\tfrac{\sqrt{3}}{2}\right) \cdot S^*(\lambda_1) \\
&= \left(C^*(\lambda_0) \oplus \tfrac{\sqrt{3}}{2} \cdot S^*(\lambda_{-1}) \right) \sim_H \left(\tfrac{1}{2} \cdot C^*(\lambda_{-1}) \oplus \tfrac{1}{2} \cdot C^*(\lambda_1) \oplus \tfrac{\sqrt{3}}{2} \cdot S^*(\lambda_1) \right)
\end{aligned}
$$

$t = 3$:

$$
x_3^* = C^*(\lambda_{-1}) \oplus_H C^*(\lambda_0) \oplus_H C^*(\lambda_1)
$$

The equation for $t = 3$ implies that all coefficients $C^*(\lambda_k), k = -1, 0, 1$, are real numbers. Therefore the following equations for the first two time points remain:

$$
x_1^* = \frac{\sqrt{3}}{2} \cdot S^*(\lambda_1) \sim_H \frac{\sqrt{3}}{2} \cdot S^*(\lambda_{-1}) \tag{24.11}
$$

$$
x_2^* = \frac{\sqrt{3}}{2} \cdot S^*(\lambda_{-1}) \sim_H \frac{\sqrt{3}}{2} \cdot S^*(\lambda_1). \tag{24.12}
$$

From $x_1^* = x_2^*$ we obtain

$$
\frac{\sqrt{3}}{2} \cdot S^*(\lambda_1) \sim_H \frac{\sqrt{3}}{2} \cdot S^*(\lambda_{-1}) = \frac{\sqrt{3}}{2} \cdot S^*(\lambda_{-1}) \sim_H \frac{\sqrt{3}}{2} \cdot S^*(\lambda_1). \tag{24.13}
$$

Equation (24.13) can only be fulfilled if $S^*(\lambda_1)$ and $S^*(\lambda_{-1})$ are real numbers. But in this case Equations (24.11) and (24.12) cannot be fulfilled.   □

Therefore for fuzzy time series the frequency approach is not used.

25

More on fuzzy random variables and fuzzy random vectors

For model-based inference methods in time series analysis the concept of fuzzy random variables is useful. Basic definitions and some propositions concerning fuzzy random variables are given in Chapter 9 (A Law of Large Numbers).

For generalized stochastic methods in time series analysis more details on fuzzy random variables are necessary. These will be explained in this section.

25.1 Basics

Definition 25.1: A *fuzzy random vector* (also called an *n-dimensional fuzzy random variable*) is a mapping X from some probability space $(\Omega, \mathcal{E}, \mathcal{P})$ into the space $(\mathcal{F}(\mathbb{R}^n), h_\infty)$ with $h_\infty(\cdot, \cdot)$ the metric defined in (23.4) if X is measurable relative to the σ-field in $\mathcal{F}(\mathbb{R}^n)$ which is generated by the open sets

$$\left\{ x^* \in \mathcal{F}(\mathbb{R}^n) : \sup_{\delta \in (0;1]} d_H\left(C_\delta(x^*), C_\delta(y^*)\right) < r \right\}$$
$$= \left\{ x^* \in \mathcal{F}(\mathbb{R}^n) : h_\infty(x^*, y^*) < r \right\},$$

with $y^* \in \mathcal{F}(\mathbb{R}^n)$ and $r > 0$.

Remark 25.1: Using $(\mathcal{F}(\mathbb{R}^n), h_\infty)$ in Definition 25.1 is important as pointed out in Körner (1997b).

Statistical Methods for Fuzzy Data   Reinhard Viertl
© 2011 John Wiley & Sons, Ltd

Theorem 25.1: Let X^* and Y^* be fuzzy random vectors, and $\delta \in (0; 1]$, and $\boldsymbol{u} \in S^{n-1}$. Then $\|X^*\|_2$, $s_{X^*}(\boldsymbol{u}, \delta)$ and $\rho_2(X^*, Y^*)$ are random variables. Moreover the Steiner point σ_{X^*} is an n-dimensional stochastic vector.

For the proof see Körner (1997b).

Definition 25.2: Two fuzzy random vectors X_1^* and X_2^* are called independent if

$$\mathcal{P}\left(X_1^* \in B_1^*, X_2^* \in B_2^*\right) = \mathcal{P}\left(X_1^* \in B_1^*\right) \mathcal{P}\left(X_2^* \in B_2^*\right),$$

for all elements B_1^* and B_2^* in the corresponding σ-fields.

Remark 25.2: In the following only fuzzy random variables with values in $\mathcal{F}_2(\mathbb{R}^n)$ are considered, i.e. $X^*(\omega) \in \mathcal{F}_2(\mathbb{R}^n)\ \forall\, \omega \in \Omega$. Such fuzzy random variables are quadratic integrable with $\mathbb{E}\,\|X^*\|_2^2 < \infty$.

25.2 Expectation and variance of fuzzy random variables

Let $\widetilde{X} : \Omega \to \mathcal{M}$ be a random quantity defined on $(\Omega, \mathcal{E}, \mathcal{P})$ with values in a metric space $(\mathcal{M}, \widetilde{\rho})$. Using the so-called *Fréchet principle* the (not necessarily unique) expectation $\mathbb{E}_F\,\widetilde{X}$ is defined to be the set of all $A \in \mathcal{M}$, which minimize the squared distance $\mathbb{E}\,\widetilde{\rho}^2\left(\widetilde{X}, A\right)$, i.e.

$$\mathbb{E}_F\,\widetilde{X} = \left\{ A \in \mathcal{M} : \mathbb{E}\,\widetilde{\rho}^2\left(\widetilde{X}, A\right) = \inf_{B \in \mathcal{M}} \mathbb{E}\,\widetilde{\rho}^2\left(\widetilde{X}, B\right) \right\}, \qquad (25.1)$$

if an $A \in \mathcal{M}$ exists with $\mathbb{E}\,\widetilde{\rho}^2\left(\widetilde{X}, A\right) < \infty$ (see Fréchet, 1948).

The infimum of the expected quadratic distance is called the *Fréchet variance*, i.e.

$$\mathrm{Var}_F\,\widetilde{X} := \inf_{A \in \mathcal{M}} \mathbb{E}\,\widetilde{\rho}^2\left(\widetilde{X}, A\right). \qquad (25.2)$$

Remark 25.3: For classical random variables X we have

$$\mathrm{Var}\,X = \mathbb{E}\,(X - \mathbb{E}\,X)^2 = \inf_{a \in \mathbb{R}} \mathbb{E}\,(X - a)^2.$$

Therefore the Fréchet principle generalizes the classical variance.

The Fréchet principle for the expectation and the variance can be applied for fuzzy random quantities X^* with values in the metric space $(\mathcal{F}_2(\mathbb{R}^n), \rho_2)$. The expectation $\mathbb{E}_F\,X^*$ is calculated as

$$\mathbb{E}_F\,X^* = \left\{ \boldsymbol{x}^* \in \mathcal{F}_2(\mathbb{R}^n) : \mathbb{E}\,\rho_2^2(X^*, \boldsymbol{x}^*) = \inf_{\boldsymbol{y}^* \in \mathcal{F}_2(\mathbb{R}^n)} \mathbb{E}\,\rho_2^2(X^*, \boldsymbol{y}^*) \right\} \qquad (25.3)$$

and the variance using (23.5) as

$$
\begin{aligned}
\mathrm{Var}_F \, X^* &= \mathbb{E} \, \rho_2^2(X^*, \mathbb{E}_F \, X^*) \\
&= \mathbb{E} \left[n \int_0^1 \int_{S^{n-1}} |s_{X^*}(u, \delta) - s_{\mathbb{E}_F X^*}(u, \delta)|^2 \, \mu(d\,u)\,d\,\delta \right] \\
&= n \int_0^1 \int_{S^{n-1}} \mathrm{Var} s_{X^*}(u, \delta) \, \mu(d\,u)\,d\,\delta.
\end{aligned}
\tag{25.4}
$$

Another possibility for the definition of the expectation is the so-called *Aumann expectation*.

Definition 25.3: The Aumann expectation $\mathbb{E}_A X^*$ of a fuzzy random variable X^* is defined by its δ-cuts $C_\delta(\mathbb{E}_A \, X^*)$, $\delta \in (0; 1]$, using families of n-dimensional random vectors $Z : \Omega \to \mathbb{R}^n$ by

$$
C_\delta(\mathbb{E}_A \, X^*) := \left\{ \mathbb{E} \, Z : Z(\omega) \in C_\delta\Big(X^*(\omega)\Big) \, \mathcal{P} \text{ a.e. and } \mathbb{E} \, \|Z\|_2 < \infty \right\}.
\tag{25.5}
$$

By the one-to-one correspondence between X^* and their δ-cuts $C_\delta(X^*)$, the Aumann expectation is clearly defined if all $C_\delta(\mathbb{E}_A \, X^*)$ for $\delta \in (0; 1]$ are non-empty.

A third possibility for the definition of the expectation of X^* is the so-called *Bochner expectation*.

Definition 25.4: The Bochner expectation $\mathbb{E} X^*$ of a fuzzy stochastic quantity X^* is defined by the support function

$$
s_{\mathbb{E} \, X^*}(u, \delta) = \mathbb{E} \, s_{X^*}(u, \delta) \qquad \forall u \in S^{n-1}, \forall \delta \in (0; 1].
\tag{25.6}
$$

Remark 25.4: The Fréchet expectation and the Fréchet variance depend on the norm used. It can be proved [see Näther (1997) and Körner and Näther (1998)], that for a fuzzy stochastic quantity X^* with values in the metric space $(\mathcal{F}(\mathbb{R}^n), h_p)$, $p \in (1; \infty)$, the Fréchet expectation

$$
\mathbb{E}_F^{(h_p)} \, X^* = \left\{ x^* \in \mathcal{F}(\mathbb{R}^n) : \mathbb{E} \, h_p^2(X^*, x^*) = \inf_{y^* \in \mathcal{F}(\mathbb{R}^n)} \mathbb{E} \, h_p^2(X^*, y^*) \right\}
$$

is in general not linear, i.e.

$$
\mathbb{E}_F^{(h_p)} \, (X^* \oplus Y^*) \neq \mathbb{E}_F^{(h_p)} \, X^* \oplus \mathbb{E}_F^{(h_p)} \, Y^*
$$

is possible.

Contrary to that, the Aumann expectation is linear:

$$
\mathbb{E}_A \, (\lambda_1 \cdot X^* \oplus \lambda_2 \cdot Y^*) = \lambda_1 \cdot \mathbb{E}_A \, X^* \oplus \lambda_2 \cdot \mathbb{E}_A \, Y^*.
$$

Moreover in Körner (1997b) the following is proved: For the norm $\rho_2(\cdot, \cdot)$ the Aumann expectation and the Bochner expectation are identical, and are uniquely determined. For this norm the Fréchet expectation is linear. Therefore in the following the Fréchet expectation is used for $\mathbb{E}\, X^*$, and the Fréchet variance for $\mathrm{Var}X^*$, based on fuzzy random quantities X^*.

Remark 25.5: By definition the Fréchet variance is a real number. There are other definitions for the generalization of the variance to fuzzy random quantities. A survey concerning different definitions for the variance of a fuzzy random quantity X^* is given in Körner (1997a). Natural conditions for the variance of a fuzzy stochastic quantity X^* are the following:

(1) If X^* is constant almost everywhere, then its variance is equal to 0:

$$\mathrm{Var}X^* = 0 \quad \Leftrightarrow \quad \mathcal{P}\{\omega \in \Omega : X^*(\omega) = x^*\} = 1 \quad \text{for } x^* \in \mathcal{F}_2\left(\mathbb{R}^n\right).$$

(2) For all $\lambda \in \mathbb{R}$ it follows

$$\mathrm{Var}(\lambda \cdot X^*) = \lambda^2 \, \mathrm{Var}(X^*).$$

(3) The variance does not depend on the location:

$$\mathrm{Var}(X^* \oplus x^*) = \mathrm{Var}(X^*) \qquad \forall x^* \in \mathcal{F}_2\left(\mathbb{R}^n\right).$$

Lemma 25.1: Let X^* and Y^* be fuzzy random variables, and $X : \Omega \to \mathbb{R}$, $x^* \in \mathcal{F}_2\left(\mathbb{R}^n\right)$, and $\lambda, \gamma \in \mathbb{R}$. Then the following holds:

(1) $\mathbb{E}\left(\lambda \cdot X^* \oplus \gamma \cdot Y^*\right) = \lambda \cdot \mathbb{E}\, X^* \oplus \gamma \cdot \mathbb{E}\, Y^*$

(2) $\mathrm{Var}X^* = \mathbb{E}\,\|X^*\|_2^2 - \|\mathbb{E}\, X^*\|_2^2$

(3) $\mathrm{Var}(\lambda \cdot X^*) = \lambda^2 \, \mathrm{Var}X^*$

(4) $\mathrm{Var}(X \cdot x^*) = \|x^*\|_2^2 \, \mathrm{Var}X$

(5) $\mathrm{Var}(X^* \oplus x^*) = \mathrm{Var}X^*$

(6) $\mathrm{Var}(X^* \oplus Y^*) = \mathrm{Var}X^* + \mathrm{Var}Y^*$ if X^* and Y^* are independent.

For the proof see Körner (1997a).

Expectation and variance of X^* can be decomposed into a local part and a fuzzy part:

$$\mathbb{E}\, X^* = \mathbb{E}\, X_0^* \oplus \mathbb{E}\, \sigma_{X^*}$$
$$\mathrm{Var}X^* = \mathrm{Var}X_0^* + \mathrm{Var}\sigma_{X^*} \tag{25.7}$$

where $X_0^* = X^* \ominus \sigma_{X^*}$ is the centered quantity. The first equation follows from the linearity of the expectation operator (cf. Lemma 25.1), and the second from the definition of the variance (25.4) and formula (23.8).

Remark 25.6: The variance is a non-negative real number. For the special case of a classical random vector $X : \Omega \to \mathbb{R}^n$ it cannot be a generalization of the $n \times n$-dimensional variance–covariance matrix. Using the notation u^T for the transposed vector of u, and I_n for the n-dimensional unit matrix, the following holds:

$$s_{X(\omega)}(u, \delta) = \langle u, X(\omega) \rangle = u^T X(\omega), \quad \omega \in \Omega,$$
$$s_{\mathbb{E} X}(u, \delta) = \langle u, \mathbb{E} X \rangle = u^T \mathbb{E} X$$

and

$$\mathrm{Var} X = \mathbb{E}\left[n \int_0^1 \int_{S^{n-1}} |s_X(u, \delta) - s_{\mathbb{E} X}(u, \delta)|^2 \, \mu(d u) \, d\delta \right]$$
$$= \mathbb{E}\left[n \int_0^1 \int_{S^{n-1}} |u^T X - u^T \mathbb{E} X|^2 \, \mu(d u) \, d\delta \right]$$
$$= \mathbb{E}\left[n \int_0^1 \int_{S^{n-1}} (X - \mathbb{E} X)^T u \, u^T (X - \mathbb{E} X) \, \mu(d u) \, d\delta \right]$$
$$= \mathbb{E}\left[(X - \mathbb{E} X)^T \underbrace{\left(n \int_0^1 \int_{S^{n-1}} u \, u^T \, \mu(d u) \, d\delta \right)}_{I_n} (X - \mathbb{E} X) \right]$$
$$= \mathbb{E}(X - \mathbb{E} X)^T (X - \mathbb{E} X).$$

For classical stochastic vectors X the Fréchet variance is the trace of its variance–covariance matrix $\mathbb{E}(X - \mathbb{E} X)(X - \mathbb{E} X)^T$. For classical (one-dimensional) random variables it is equal to its standard variance.

25.3   Covariance and correlation

Definition 25.5: For two fuzzy random variables X^* and Y^* the covariance $\mathrm{Cov}(X^*, Y^*)$ is defined using the scalar product of the support functions from (23.9):

$$\mathrm{Cov}(X^*, Y^*) := \mathbb{E}\langle s_{X^*} - s_{\mathbb{E} X^*}, s_{Y^*} - s_{\mathbb{E} Y^*} \rangle$$
$$= \mathbb{E}\left[n \int_0^1 \int_{S^{n-1}} (s_{X^*} - s_{\mathbb{E} X^*})(s_{Y^*} - s_{\mathbb{E} Y^*}) \, \mu(d u) \, d\delta \right]$$
$$= n \int_0^1 \int_{S^{n-1}} \mathrm{Cov}(s_{X^*}, s_{Y^*}) \, \mu(d u) \, d\delta.$$

The correlation $\text{Corr}(X^*, Y^*)$ is defined by

$$\text{Corr}(X^*, Y^*) := \frac{\text{Cov}(X^*, Y^*)}{\sqrt{\text{Var}(X^*)}\sqrt{\text{Var}(Y^*)}}.$$

This covariance can be decomposed into a local and a fuzzy part. With $X_0^* = X^* \ominus \sigma_{X^*}$ and $Y_0^* = Y^* \ominus \sigma_{Y^*}$ we obtain

$$\text{Cov}(X^*, Y^*) = \text{Cov}(X_0^*, Y_0^*) + \text{Cov}(\sigma_{X^*}, \sigma_{Y^*}).$$

This is proved by using the definition of the covariance.

Lemma 25.2: Covariance and correlation of two fuzzy random quantities X^* and Y^* fulfill the following:

(1) $\text{Cov}(X^*, X^*) = \text{Var}X^*$

(2) $\text{Var}(X^* \oplus Y^*) = \text{Var}X^* + \text{Var}Y^* + 2\,\text{Cov}(X^*, Y^*)$

(3) $\text{Corr}(X^*, Y^*) = 0$ if X^* and Y^* are stochastically independent

(4) $|\text{Corr}(X^*, Y^*)| \leq 1$

(5) $\text{Cov}(X^*, Y^*) = \mathbb{E}\langle X^*, Y^* \rangle - \langle \mathbb{E}\,X^*, \mathbb{E}\,Y^* \rangle.$

For the proof see Körner (1997a).
The following proposition gives further properties of covariance and correlation.

Proposition 25.1: For fuzzy random quantities X^*, X_1^*, X_2^* and Y^*, and $x^*, y^* \in \mathcal{F}_2(\mathbb{R}^n)$ the following holds:

(1) $\text{Cov}(X_1^* \oplus X_2^*, Y^*) = \text{Cov}(X_1^*, Y^*) + \text{Cov}(X_2^*, Y^*)$

(2) $\mathbb{E}\langle X^*, x^* \rangle = \langle \mathbb{E}\,X^*, x^* \rangle$

(3) $\text{Cov}(\lambda \cdot X^* \oplus x^*, \nu \cdot Y^* \oplus y^*) = \lambda\,\nu\,\text{Cov}(X^*, Y^*)$ for $\lambda, \nu \in \mathbb{R}$ with $\lambda\,\nu \geq 0$

(4) $\text{Corr}(X^*, Y^*) = 1 \quad \Leftrightarrow \quad Y^* \oplus y^* = \alpha \cdot X^* \oplus x^*$ with $\alpha \geq 0$

(5) $\text{Corr}(X^*, Y^*) = -1 \quad \Leftrightarrow \quad Y^* \oplus \alpha \cdot X^* = y^*$ with $\alpha \geq 0.$

Proof:

ad (1) This follows immediately from the definition of the covariance and the properties of the support function.

ad (2) This follows directly from the definition of the scalar product [cf. (23.9) and (25.6)].

ad (3) By Lemma 25.1 and (23.7) for $\lambda\,\nu \geq 0$, i.e. $\operatorname{sign}(\lambda) = \operatorname{sign}(\nu)$ we obtain:

$$
\begin{aligned}
&\operatorname{Cov}\left(\lambda \cdot X^* \oplus x^*, \nu \cdot Y^* \oplus y^*\right) \\
&= \mathbb{E}\left\langle s_{\lambda \cdot X^* \oplus x^*} - s_{\lambda \cdot \mathbb{E} X^* \oplus x^*},\ s_{\nu \cdot Y^* \oplus y^*} - s_{\nu \cdot \mathbb{E} Y^* \oplus y^*}\right\rangle \\
&= \mathbb{E}\left\langle s_{\lambda \cdot X^*} + s_{x^*} - s_{\lambda \cdot \mathbb{E} X^*} - s_{x^*},\ s_{\nu \cdot Y^*} + s_{y^*} - s_{\nu \cdot \mathbb{E} Y^*} - s_{y^*}\right\rangle \\
&= \mathbb{E}\left\langle |\lambda|\, s_{\operatorname{sign}(\lambda)\cdot X^*} - |\lambda|\, s_{\operatorname{sign}(\lambda)\cdot \mathbb{E} X^*},\ |\nu|\, s_{\operatorname{sign}(\nu)\cdot Y^*} - |\nu|\, s_{\operatorname{sign}(\nu)\cdot \mathbb{E} Y^*}\right\rangle \\
&= |\lambda|\,|\nu|\,\mathbb{E}\left\langle s_{\operatorname{sign}(\lambda)\cdot X^*} - s_{\operatorname{sign}(\lambda)\cdot \mathbb{E} X^*},\ s_{\operatorname{sign}(\nu)\cdot Y^*} - s_{\operatorname{sign}(\nu)\cdot \mathbb{E} Y^*}\right\rangle \\
&= \lambda\,\nu\,\operatorname{Cov}\left(X^*, Y^*\right).
\end{aligned}
$$

ad (4) The necessity of the condition is seen in the following way: Defining $\beta = \sqrt{\dfrac{\operatorname{Var}Y^*}{\operatorname{Var}X^*}}$, $\alpha = \beta$, $y^* = \beta \cdot \mathbb{E} X^*$ and $x^* = \mathbb{E} Y^*$ we have $\operatorname{Corr}(X^*, Y^*) = 1 \Rightarrow \operatorname{Cov}(X^*, Y^*) = \sqrt{\operatorname{Var}X^* \operatorname{Var}Y^*}$ and by condition (2) for the quadratic distance of the fuzzy random quantities $Y^* \oplus y^*$ and $\alpha \cdot X^* \oplus x^*$ we obtain

$$
\begin{aligned}
&\mathbb{E}\,\rho_2^2\left(Y^* \oplus y^*, \alpha \cdot X^* \oplus x^*\right) = \mathbb{E}\,\rho_2^2\left(Y^* \oplus \beta \cdot \mathbb{E} X^*, \beta \cdot X^* \oplus \mathbb{E} Y^*\right) \\
&= \mathbb{E}\left\langle Y^* \oplus \beta \cdot \mathbb{E} X^*, Y^* \oplus \beta \cdot \mathbb{E} X^*\right\rangle - 2\,\mathbb{E}\left\langle Y^* \oplus \beta \cdot \mathbb{E} X^*, \beta \cdot X^* \oplus \mathbb{E} Y^*\right\rangle \\
&\quad + \mathbb{E}\left\langle \beta \cdot X^* \oplus \mathbb{E} Y^*, \beta \cdot X^* \oplus \mathbb{E} Y^*\right\rangle \\
&= \mathbb{E}\left\langle Y^*, Y^*\right\rangle + 2\,\beta\,\mathbb{E}\left\langle Y^*, \mathbb{E} X^*\right\rangle + \beta^2\left\langle \mathbb{E} X^*, \mathbb{E} X^*\right\rangle - 2\,\mathbb{E}\left\langle Y^*, \mathbb{E} Y^*\right\rangle \\
&\quad - 2\,\beta\,\mathbb{E}\left\langle Y^*, X^*\right\rangle - 2\,\beta\left\langle \mathbb{E} X^*, \mathbb{E} Y^*\right\rangle - 2\,\beta^2\,\mathbb{E}\left\langle \mathbb{E} X^*, X^*\right\rangle \\
&\quad + \beta^2\,\mathbb{E}\left\langle X^*, X^*\right\rangle + 2\,\beta\,\mathbb{E}\left\langle X^*, \mathbb{E} Y^*\right\rangle + \left\langle \mathbb{E} Y^*, \mathbb{E} Y^*\right\rangle \\
&= \mathbb{E}\left\langle Y^*, Y^*\right\rangle + 2\,\beta\left\langle \mathbb{E} Y^*, \mathbb{E} X^*\right\rangle + \beta^2\left\langle \mathbb{E} X^*, \mathbb{E} X^*\right\rangle - 2\left\langle \mathbb{E} Y^*, \mathbb{E} Y^*\right\rangle \\
&\quad - 2\,\beta\,\mathbb{E}\left\langle Y^*, X^*\right\rangle - 2\,\beta\left\langle \mathbb{E} X^*, \mathbb{E} Y^*\right\rangle - 2\,\beta^2\left\langle \mathbb{E} X^*, \mathbb{E} X^*\right\rangle \\
&\quad + \beta^2\,\mathbb{E}\left\langle X^*, X^*\right\rangle + 2\,\beta\left\langle \mathbb{E} X^*, \mathbb{E} Y^*\right\rangle + \left\langle \mathbb{E} Y^*, \mathbb{E} Y^*\right\rangle \\
&= \mathbb{E}\left\langle Y^*, Y^*\right\rangle - \left\langle \mathbb{E} Y^*, \mathbb{E} Y^*\right\rangle - \left(2\,\beta\,\mathbb{E}\left\langle Y^*, X^*\right\rangle - 2\,\beta\left\langle \mathbb{E} X^*, \mathbb{E} Y^*\right\rangle\right) \\
&\quad + \beta^2\,\mathbb{E}\left\langle X^*, X^*\right\rangle - \beta^2\left\langle \mathbb{E} X^*, \mathbb{E} X^*\right\rangle \\
&= \operatorname{Var}Y^* - 2\,\beta\,\operatorname{Cov}\left(X^*, Y^*\right) + \beta^2\,\operatorname{Var}X^* \\
&= \operatorname{Var}Y^* - 2\sqrt{\frac{\operatorname{Var}Y^*}{\operatorname{Var}X^*}}\sqrt{\operatorname{Var}X^* \operatorname{Var}Y^*} + \frac{\operatorname{Var}Y^*}{\operatorname{Var}X^*}\operatorname{Var}X^* = 0.
\end{aligned}
$$

The fuzzy random quantities $Y^* \oplus y^*$ and $\alpha \cdot X^* \oplus x^*$ are identical $\mathcal{P}$ – almost everywhere.

The sufficiency of the condition follows from (3) and conditions (3) and (5) from Lemma 25.1:

$$\text{Cov}\,(X^*, Y^*) = \text{Cov}\,(X^*, Y^* \oplus y^*) = \text{Cov}\,(X^*, \alpha \cdot X^* \oplus x^*)$$
$$= \alpha \,\text{Cov}\,(X^*, X^*) = \sqrt{\alpha^2 \,\text{Var}X^* \,\text{Var}X^*}$$
$$= \sqrt{\text{Var}X^* \,\text{Var}(\alpha \cdot X^* \oplus x^*)} = \sqrt{\text{Var}X^* \,\text{Var}(Y^* \oplus y^*)}$$
$$1 = \sqrt{\text{Var}X^* \,\text{Var}Y^*}.$$

ad (5) The necessity follows similar to the proof of the necessity in (4): From $\beta = \sqrt{\frac{\text{Var}Y^*}{\text{Var}X^*}}$, $\alpha = \beta$ and $y^* = \beta \cdot \mathbb{E}\,X^* \oplus \mathbb{E}\,Y^*$ it follows $\mathbb{E}\,\rho_2^2(Y^* \oplus \alpha \cdot X^*, y^*) = \mathbb{E}\,\rho_2^2(Y^* \oplus \beta \cdot X^*, \beta \cdot \mathbb{E}\,X^* \oplus \mathbb{E}\,Y^*) = 0$.

For the sufficiency, from (1) and (3) we obtain

$$\text{Cov}\,(X^*, Y^*) = \text{Cov}\,(X^* \oplus y^*, Y^*) = \text{Cov}\,(X^* \oplus Y^* \oplus \alpha \cdot X^*, Y^*)$$
$$= (1 + \alpha)\,\text{Cov}\,(X^*, Y^*) + \text{Var}Y^*, \text{ from which follows}$$
$$-\alpha \,\text{Cov}\,(X^*, Y^*) = \text{Var}Y^* \qquad (25.8)$$

and

$$\text{Cov}\,(X^*, Y^*) = \text{Cov}\,(X^*, Y^* \oplus y^*) = \text{Cov}\,(X^*, Y^* \oplus Y^* \oplus \alpha \cdot X^*)$$
$$= 2\,\text{Cov}\,(X^*, Y^*) + \alpha \,\text{Var}X^* \quad \Rightarrow$$
$$-\text{Cov}\,(X^*, Y^*) = \alpha \,\text{Var}X^*. \qquad (25.9)$$

From Equations (25.8) and (25.9) we obtain $\text{Var}Y^* = \alpha^2 \,\text{Var}X^*$ and from Equation (25.9)

$$-\text{Cov}\,(X^*, Y^*) = \alpha \,\text{Var}X^* = \sqrt{\alpha^2 \,\text{Var}X^* \,\text{Var}X^*} = \sqrt{\text{Var}Y^* \,\text{Var}X^*}.$$

□

25.4 Further results

A fuzzy stochastic quantity X^* in LR-form is defined by four real random variables m, s, l and r with $\mathcal{P}\{s \geq 0\} = \mathcal{P}\{l \geq 0\} = \mathcal{P}\{r \geq 0\} = 1$ and two fuzzy numbers $x_L^*, x_R^* \in \mathcal{F}_2(\mathbb{R})$:

$$X^* = \langle m, s, l, r \rangle_{LR} := m \oplus s \cdot [-1, 1] \ominus l \cdot x_L^* \oplus r \cdot x_R^*.$$

In order to fulfill $\mathbb{E}\,\|X^*\|_2^2$ the random variables m, s, l and r have to be square-integrable. By the linearity of the expectation operator the expectation $\mathbb{E}\,X^*$ is a fuzzy number in LR-form obeying

$$\mathbb{E}\,X^* = \langle \mathbb{E}\,m, \mathbb{E}\,s, \mathbb{E}\,l, \mathbb{E}\,r \rangle_{LR} = \mathbb{E}\,m \oplus \mathbb{E}\,s \cdot [-1, 1] \ominus \mathbb{E}\,l \cdot x_L^* \oplus \mathbb{E}\,r \cdot x_R^*.$$

The variance of this fuzzy random quantity is given by

$$\operatorname{Var} X^* = \operatorname{Var} m + \operatorname{Var} s + \|x_L^*\|_2^2 \operatorname{Var} l + \|x_R^*\|_2^2 \operatorname{Var} r$$
$$- 2 \|x_L^*\|_1 \Big[\operatorname{Cov}(m, l) - \operatorname{Cov}(s, l) \Big] + 2 \|x_R^*\|_1 \Big[\operatorname{Cov}(m, r) - \operatorname{Cov}(s, r) \Big].$$

Many results for real random variables can be generalized for fuzzy random quantities.

Proposition 25.2 (Chebysheff's inequality): For fuzzy stochastic quantities X^* with $\mathbb{E} \|X^*\|_2^2 < \infty$ the following holds:

$$P\Big(\rho_2(X^*, \mathbb{E} X^*) \geq \varepsilon\Big) \leq \frac{\operatorname{Var}(X^*)}{\varepsilon^2} \qquad \forall \varepsilon > 0.$$

Proof: By the definition of the variance we have

$$\frac{\operatorname{Var}(X^*)}{\varepsilon^2} = \frac{1}{\varepsilon^2} \int_\Omega \rho_2^2 \Big(X^*(\omega), \mathbb{E} X^*\Big) P(d\omega)$$
$$\geq \frac{1}{\varepsilon^2} \int_\Omega I_{\{\omega : \rho_2^2(X^*(\omega), \mathbb{E} X^*) \geq \varepsilon^2\}}(\omega)\, \rho_2^2 \Big(X^*(\omega), \mathbb{E} X^*\Big) P(d\omega)$$
$$\geq \int_\Omega I_{\{\omega : \rho_2^2(X^*(\omega), \mathbb{E} X^*) \geq \varepsilon^2\}}(\omega)\, P(d\omega) = P\Big(\rho_2(X^*, \mathbb{E} X^*) \geq \varepsilon\Big).$$

$\square$

The generalized Chebysheff inequality can be used to prove a generalized weak law of large numbers.

Theorem 25.2 (Weak law of large numbers): Let $(X_n^*)_{n \in \mathbb{N}}$ be a sequence of independent and identical distributed fuzzy random quantities obeying $\mathbb{E} \|X_1^*\|_2^2 < \infty$. Defining $X^*(n) = \dfrac{1}{n} \cdot \bigoplus_{i=1}^n X_i^*$ the following holds:

$$\lim_{n \to \infty} P\Big(\rho_2 \big(X^*(n), \mathbb{E} X_1^*\big) \geq \varepsilon\Big) = 0 \qquad \forall \varepsilon > 0.$$

Proof: The proof is analogous to the classical theorem obeying $\operatorname{Var} X^*(n) = n^{-1} \operatorname{Var} X_1^*$ which follows from Lemma 25.1.

Theorem 25.3 (Kolmogorov's inequality): For independent fuzzy random quantities $X_1^*, \ldots, X_n^*$ obeying $\mathbb{E} \|X_i^*\|_2^2 < \infty$ for $i = 1 \,(1)\, n$ the following inequality holds:

$$P\left(\max_{1 \leq k \leq n} \rho_2 \left(\bigoplus_{i=1}^k X_i^*, \bigoplus_{i=1}^k \mathbb{E} X_i^* \right) \geq \varepsilon \right) \leq \frac{1}{\varepsilon^2} \sum_{i=1}^n \operatorname{Var} X_i^* \qquad \forall \varepsilon > 0.$$

For the proof see Körner (1997a).

From the Kolmogorov inequality we obtain a generalized strong law of large numbers.

Theorem 25.4 (Strong law of large numbers): For a sequence $(X_n^*)_{n \in \mathbb{N}}$ of independent fuzzy random quantities with convergent sequence $\sum\limits_{n=1}^{\infty} \dfrac{1}{n^2} \operatorname{Var} X_n^*$ and $\mathbb{E} \, \|X_n^*\|_2^2 < \infty$ for $n \in \mathbb{N}$ the following holds:

$$\mathcal{P}\left(\lim_{n \to \infty} \rho_2 \left(\frac{1}{n} \cdot \bigoplus_{i=1}^{n} X_i^*, \, \frac{1}{n} \cdot \bigoplus_{i=1}^{n} \mathbb{E} \, X_i^*\right) = 0\right) = 1$$

For the proof see Körner (1997a).

26

Stochastic methods in fuzzy time series analysis

In Chapter 24 fuzzy time series are analyzed without assumptions on a stochastic model. The only stochastic element was the error component ε_t^* for moving averages.

In the present chapter the fuzzy observations are considered as realizations of a stochastic process whose elements are fuzzy stochastic quantities. The goal is to use dependencies of the observations for predictions of future values.

26.1 Linear approximation and prediction

A fuzzy stochastic process $(X_t^*)_{t \in T}$ is a family of fuzzy random quantities X_t^*, where T is an index set, usually a subset of the set of real numbers $\mathbb{R}$. The first approach is looking for the 'best' linear approximation of the fuzzy random quantity X_t^*, $t \in T$, from the p quantities $X_{t-1}^*, \ldots, X_{t-p}^*$, $p > 0$ with $t > p$. The approximation is assumed to be

$$X_t^* \approx \bigoplus_{i=1}^{p} \alpha_i \cdot X_{t-i}^* \tag{26.1}$$

with real numbers α_i. The quality of the approximation is determined by the expectation of the squared distance between approximated value and observed value. This means for the coefficients α_i looking for those values $(\widehat{\alpha}_1, \ldots, \widehat{\alpha}_p) \in \mathbb{R}^p$ fulfilling

$$\mathbb{E}\,\rho_2^2\left(X_t^*, \bigoplus_{i=1}^{p} \widehat{\alpha}_i \cdot X_{t-i}^*\right) = \min_{(\alpha_1, \ldots, \alpha_p) \in \mathbb{R}^p} \mathbb{E}\,\rho_2^2\left(X_t^*, \bigoplus_{i=1}^{p} \alpha_i \cdot X_{t-i}^*\right). \tag{26.2}$$

Statistical Methods for Fuzzy Data   Reinhard Viertl
© 2011 John Wiley & Sons, Ltd

In the case of real valued weakly stationary time series $(X_t)_{t\in T}$, i.e. constant expectation $\mathbb{E}\,X_t = \mu$ for all $t \in T$ and covariance $\mathrm{Cov}\,(X_t, X_s)$ depending only on the distance $t - s$ for all $t, s \in T$ for $\mathbb{E}\,X_t = 0$ and $\mathrm{Cov}\,(X_t, X_s) = \gamma_{t-s}$ the coefficients α_i, $i = 1\,(1)\,p$, can be obtained from the so-called Yule–Walker equations of order p. For details see Schlittgen and Streitberg (1984).

From

$$
\mathbb{E}\left(X_t - \sum_{i=1}^{p}\widehat{\alpha}_i\,X_{t-i}\right)^2 = \min_{(\alpha_1,\ldots,\alpha_p)\in\mathbb{R}^p}\mathbb{E}\left(X_t - \sum_{i=1}^{p}\alpha_i\,X_{t-i}\right)^2, \tag{26.3}
$$

assuming the following matrix to be invertible, it follows:

$$
\begin{pmatrix}\widehat{\alpha}_1 \\ \widehat{\alpha}_2 \\ \vdots \\ \widehat{\alpha}_{p-1} \\ \widehat{\alpha}_p\end{pmatrix} =
\begin{pmatrix}
\gamma_0 & \gamma_1 & \gamma_2 & \cdots & \gamma_{p-2} & \gamma_{p-1} \\
\gamma_1 & \gamma_0 & \gamma_1 & \cdots & \gamma_{p-3} & \gamma_{p-2} \\
\vdots & \vdots & \vdots & \ddots & \vdots & \vdots \\
\gamma_{p-2} & \gamma_{p-3} & \gamma_{p-4} & \cdots & \gamma_2 & \gamma_1 \\
\gamma_{p-1} & \gamma_{p-2} & \gamma_{p-3} & \cdots & \gamma_1 & \gamma_0
\end{pmatrix}^{-1}
\begin{pmatrix}\gamma_1 \\ \gamma_2 \\ \vdots \\ \gamma_{p-1} \\ \gamma_p\end{pmatrix}
$$

In the case of fuzzy data, by Equation (26.2), Equation (23.10) is equivalent to

$$
\mathbb{E}\,\rho_2^2\left(X_t^*, \bigoplus_{i=1}^{p}\alpha_i \cdot X_{t-i}^*\right) = \mathbb{E}\left(\langle X_t^*, X_t^*\rangle - 2\left\langle X_t^*, \bigoplus_{i=1}^{p}\alpha_i \cdot X_{t-i}^*\right\rangle\right.
$$
$$
\left. + \left\langle \bigoplus_{i=1}^{p}\alpha_i \cdot X_{t-i}^*, \bigoplus_{i=1}^{p}\alpha_i \cdot X_{t-i}^*\right\rangle\right) \to \min.
$$

The determination of the coefficients is complicated because the separation of the scalar product is only possible for positive coefficients, which follows from Theorem 23.2.

Assuming all coefficients α_i to be positive, from Theorem 23.2 and Lemma 25.1 we obtain:

$$
\mathbb{E}\left(\langle X_t^*, X_t^*\rangle - 2\left\langle X_t^*, \bigoplus_{i=1}^{p}\alpha_i \cdot X_{t-i}^*\right\rangle + \left\langle \bigoplus_{i=1}^{p}\alpha_i \cdot X_{t-i}^*, \bigoplus_{i=1}^{p}\alpha_i \cdot X_{t-i}^*\right\rangle\right) =
$$
$$
\mathbb{E}\left(\langle X_t^*, X_t^*\rangle - 2\sum_{i=1}^{p}\alpha_i\,\langle X_t^*, X_{t-i}^*\rangle + \sum_{i=1}^{p}\sum_{j=1}^{p}\alpha_i\,\alpha_j\,\langle X_{t-i}^*, X_{t-j}^*\rangle\right) =
$$
$$
\mathbb{E}\,\langle X_t^*, X_t^*\rangle - 2\sum_{i=1}^{p}\alpha_i\,\mathbb{E}\,\langle X_t^*, X_{t-i}^*\rangle + \sum_{i=1}^{p}\sum_{j=1}^{p}\alpha_i\,\alpha_j\,\mathbb{E}\,\langle X_{t-i}^*, X_{t-j}^*\rangle \to \min
$$

Using the notation

$$\alpha = (\alpha_1, \ldots, \alpha_p)$$
$$\gamma = \left(\mathbb{E} \langle X_t^*, X_{t-1}^* \rangle, \ldots, \mathbb{E} \langle X_t^*, X_{t-p}^* \rangle \right)$$

and

$$\Gamma = \begin{pmatrix} \mathbb{E} \langle X_{t-1}^*, X_{t-1}^* \rangle & \mathbb{E} \langle X_{t-1}^*, X_{t-2}^* \rangle & \cdots & \mathbb{E} \langle X_{t-1}^*, X_{t-p}^* \rangle \\ \mathbb{E} \langle X_{t-2}^*, X_{t-1}^* \rangle & \mathbb{E} \langle X_{t-2}^*, X_{t-2}^* \rangle & \cdots & \mathbb{E} \langle X_{t-2}^*, X_{t-p}^* \rangle \\ \vdots & \vdots & \ddots & \vdots \\ \mathbb{E} \langle X_{t-p}^*, X_{t-1}^* \rangle & \mathbb{E} \langle X_{t-p}^*, X_{t-2}^* \rangle & \cdots & \mathbb{E} \langle X_{t-p}^*, X_{t-p}^* \rangle \end{pmatrix}$$

where Γ is a symmetric matrix, the problem is a minimization problem of a quadratic form

$$\mathbb{E} \rho_2^2 \left(X_t^*, \bigoplus_{i=1}^{p} \alpha_i \cdot X_{t-i}^* \right) = \alpha \, \Gamma \, \alpha^T - 2 \gamma \, \alpha^T \rightarrow \min$$

with side conditions $\alpha_i \geq 0, i = 1\,(1)\,p$.

This problem is well known in economics. For details see Luptáčik (1981).

Remark 26.1: The matrix Γ is positive semi-definite.

Proof: By definition of the inner product $\langle \cdot, \cdot \rangle$ from Theorem 23.2 for the elements in the matrix Γ the following holds:

$$\mathbb{E} \langle X_{t-i}^*, X_{t-j}^* \rangle = \mathbb{E} \int_0^1 \int_{S^0} s_{X_{t-i}^*}(u, \delta)\, s_{X_{t-j}^*}(u, \delta)\, \mu(d\,u)\, d\,\delta \,, \qquad i, j = 1\,(1)\,p.$$

The functions $s_{X_{t-i}^*}(u, \delta)$ are for all $u \in \{-1, 1\}$ and $\delta \in (0; 1]$ one-dimensional random variables by Theorem 25.1. For arbitrary vector $\alpha = (\alpha_1, \ldots, \alpha_p) \in \mathbb{R}^p$ therefore

$$\alpha \, \Gamma \, \alpha^T = \sum_{i=1}^{p} \sum_{j=1}^{p} \alpha_i \, \alpha_j \, \mathbb{E} \langle X_{t-i}^*, X_{t-j}^* \rangle$$

$$= \sum_{i=1}^{p} \sum_{j=1}^{p} \alpha_i \, \alpha_j \left(\mathbb{E} \int_0^1 \int_{S^0} s_{X_{t-i}^*}(u, \delta)\, s_{X_{t-j}^*}(u, \delta)\, \mu(d\,u)\, d\,\delta \right)$$

$$= \mathbb{E} \int_0^1 \int_{S^0} \left(\sum_{i=1}^{p} \sum_{j=1}^{p} \alpha_i \, \alpha_j \, s_{X_{t-i}^*}(u, \delta)\, s_{X_{t-j}^*}(u, \delta) \right) \mu(d\,u)\, d\,\delta$$

$$= \mathbb{E} \int_0^1 \int_{S^0} \left(\sum_{i=1}^{p} \alpha_i \, s_{X_{t-i}^*}(u, \delta) \right)^2 \mu(d\,u)\, d\,\delta \geq 0.$$

In applications the elements of the matrix Γ are not known and have to be estimated based on the data $(x_t^*)_{t \in T}$. In order to keep the number of quantities which have to be estimated reasonable, the following conditions are assumed for the fuzzy stochastic quantities $(X_t^*)_{t \in T}$:

(S1) $\mathbb{E} X_t^* = \mu^*$ $\forall t \in T$

(S2) $\sigma_{\mathbb{E} X_t^*} = \sigma_{\mu^*} = 0$

(S3) $\text{Cov} \left(X_t^*, X_s^* \right) = \gamma(t - s)$

which means the expectation is constant with Steiner point 0, and the covariance of two different fuzzy stochastic quantities X_t^* and X_s^* depends on the time difference $t - s$ only.

A more stringent condition instead of condition (S3) would be the following: For all $\delta \in (0; 1]$, $u \in \{-1, 1\}$ and any pair of one-dimensional fuzzy random quantities $X_t^*(u, \delta)$ and $X_s^*(u, \delta)$ the covariance has to fulfill $\text{Cov} \left(X_t^*(u, \delta), X_s^*(u, \delta) \right) = \gamma_{t-s}(u, \delta)$. In the following the weaker condition (S3) is used.

By the definition of the covariance it follows $\text{Cov}(X_t^*, X_s^*) = \mathbb{E} \langle X_t^*, X_s^* \rangle + \langle \mathbb{E} X_t^*, \mathbb{E} X_s^* \rangle$, also the scalar product

$$\eta(t - s) = \mathbb{E} \langle X_t^*, X_s^* \rangle = \text{Cov}(X_t^*, X_s^*) - \langle \mathbb{E} X_t^*, \mathbb{E} X_s^* \rangle = \gamma(t - s) - \langle \mu^*, \mu^* \rangle$$

depends only on the time difference $t - s$.

For classical time series $(X_t)_{t \in T}$ the above conditions correspond to the conditions $\mathbb{E} X_t = 0$ and $\text{Cov}(X_t, X_s) = \gamma(t - s)$, i.e. the conditions of a weakly stationary process. In order to estimate the matrix Γ estimates of $\mathbb{E} \langle X_t^*, X_s^* \rangle$ are needed. First, the estimation of the expectation is considered.

Similar to the standard real valued case the expectation μ^* is estimated by a temporal mean values of the observations:

$$\widehat{\mu}^* = \overline{X_N^*} = \frac{1}{N} \cdot \bigoplus_{t=1}^{N} X_t^*$$

By the linearity of the expectation operator (cf. Lemma 25.1), the mean is an unbiased estimator for μ^*:

$$\mathbb{E} \overline{X_N^*} = E \left(\frac{1}{N} \cdot \bigoplus_{t=1}^{N} X_t^* \right) = \frac{1}{N} \cdot \bigoplus_{t=1}^{N} \mathbb{E} X_t^* = \mu^*$$

For the variance of this estimator we obtain

$$\text{Var} \overline{X_N^*} = \text{Var} \left(\frac{1}{N} \cdot \bigoplus_{t=1}^{N} X_t^* \right) = \frac{1}{N^2} \sum_{i=1}^{N} \sum_{j=1}^{N} \text{Cov} \left(X_i^*, X_j^* \right) = \frac{1}{N^2} \sum_{i=1}^{N} \sum_{j=1}^{N} \gamma(i - j)$$

$$= \frac{1}{N^2} \sum_{|h|<N} (N - |h|) \, \gamma(h) = \frac{1}{N} \sum_{|h|<N} \left(1 - \frac{|h|}{N} \right) \gamma(h).$$

For $N \to \infty$ and absolute summable sequence $(\gamma(h))_{h \in \mathbb{N}}$ the sequence $\overline{X_N^*}$ is consistent for μ^*. This follows from the following inequality:

$$\operatorname{Var}\overline{X_N^*} = \frac{1}{N} \sum_{h=-(N-1)}^{N-1} \left(1 - \frac{|h|}{N}\right) \gamma(h) \leq \frac{1}{N} \sum_{h=-(N-1)}^{N-1} \left(1 - \frac{|h|}{N}\right) |\gamma(h)|$$

$$\leq \frac{1}{N} \sum_{h=-(N-1)}^{N-1} |\gamma(h)| \leq \frac{1}{N} \sum_{h=-\infty}^{\infty} |\gamma(h)|$$

An estimation of the scalar product $\eta(t - s) = \mathbb{E} \langle X_t^*, X_s^* \rangle$ using the notation $h = |t - s|$ is given by

$$\tilde{\eta}(h) = \frac{1}{N-h} \sum_{i=1}^{N-h} \langle X_i^*, X_{i+h}^* \rangle = \frac{1}{N - |t-s|} \sum_{i=\max\{1,1-(t-s)\}}^{\min\{N,N-(t-s)\}} \langle X_i^*, X_{i+(t-s)}^* \rangle. \quad (26.4)$$

This estimation is unbiased for $\eta(h)$:

$$\mathbb{E}\tilde{\eta}(h) = \mathbb{E} \left(\frac{1}{N-h} \sum_{i=1}^{N-h} \langle X_i^*, X_{i+h}^* \rangle \right) = \frac{1}{N-h} \sum_{i=1}^{N-h} \mathbb{E} \langle X_i^*, X_{i+h}^* \rangle = \eta(h)$$

In order to estimate the elements of the matrix Γ the following estimation is used:

$$\hat{\eta}(h) = \frac{1}{N} \sum_{i=1}^{N-h} \langle X_i^*, X_{i+h}^* \rangle = \frac{1}{N} \sum_{i=\max\{1,1-(t-s)\}}^{\min\{N,N-(t-s)\}} \langle X_i^*, X_{i+(t-s)}^* \rangle \quad (26.5)$$

This estimation is biased, but asymptotically unbiased, which is seen from

$$\mathbb{E}\hat{\eta}(h) = \mathbb{E} \left(\frac{1}{N} \sum_{i=1}^{N-h} \langle X_i^*, X_{i+h}^* \rangle \right) = \frac{1}{N} \sum_{i=1}^{N-h} \mathbb{E} \langle X_i^*, X_{i+h}^* \rangle = \frac{N-h}{N} \eta(h).$$

The reason for using this estimator is the positive semi-definiteness of the corresponding matrix $\hat{\Gamma}$ with

$$\hat{\Gamma} = \begin{pmatrix} \hat{\eta}(0) & \hat{\eta}(1) & \hat{\eta}(2) & \cdots & \hat{\eta}(p-2) & \hat{\eta}(p-1) \\ \hat{\eta}(1) & \hat{\eta}(0) & \hat{\eta}(1) & \cdots & \hat{\eta}(p-3) & \hat{\eta}(p-2) \\ \vdots & \vdots & \vdots & \ddots & \vdots & \vdots \\ \hat{\eta}(p-2) & \hat{\eta}(p-3) & \hat{\eta}(p-4) & \cdots & \hat{\eta}(0) & \hat{\eta}(1) \\ \hat{\eta}(p-1) & \hat{\eta}(p-2) & \hat{\eta}(p-3) & \cdots & \hat{\eta}(1) & \hat{\eta}(0) \end{pmatrix}.$$

The proof of this is given below (Lemma 26.1). Using the unbiased estimator $\tilde{\eta}(h)$ does not guarantee the matrix $\tilde{\Gamma}$ to be positive semi-definite. An example is a real time series with 10 observations $\{-2, 2, -3, 3, -4, 4, -5, 5, -2, 2\}$ and $p = 2$ for which the determinant of $\widehat{\Gamma}$ is negative.

Lemma 26.1: The matrix $\widehat{\Gamma}$ formed by the estimators $\widehat{\eta}(h)$ is positive semi-definite.

Proof: For an arbitrary vector $\alpha = (\alpha_1, \ldots, \alpha_p) \in \mathbb{R}^p$ the following holds:

$$
\alpha \, \widehat{\Gamma} \, \alpha^T = \sum_{i=1}^{p} \sum_{j=1}^{p} \alpha_i \alpha_j \eta(|i - j|)
$$

$$
= \sum_{i=1}^{p} \sum_{j=1}^{p} \alpha_i \alpha_j \left(\frac{1}{N} \sum_{l=1}^{N-|i-j|} \langle X_l^*, X_{l+|i-j|}^* \rangle \right)
$$

$$
= \sum_{i=1}^{p} \sum_{j=1}^{p} \alpha_i \alpha_j \left(\frac{1}{N} \sum_{l=1}^{N-|i-j|} \int_0^1 \int_{S^0} s_{X_l^*}(u, \delta) s_{X_{l+|i-j|}^*}(u, \delta) \mu(du) d\delta \right)
$$

$$
= \int_0^1 \int_{S^0} \sum_{i=1}^{p} \sum_{j=1}^{p} \alpha_i \alpha_j \left(\frac{1}{N} \sum_{l=1}^{N-|i-j|} s_{X_l^*}(u, \delta) s_{X_{l+|i-j|}^*}(u, \delta) \right) \mu(du) d\delta
$$

$$
= \frac{1}{N} \int_0^1 \int_{S^0} \sum_{l=1}^{N+p-1}
$$

$$
\times \left(\sum_{i=\max\{1, l+1-N\}}^{\min\{l, p\}} \sum_{j=\max\{1, l+1-N\}}^{\min\{l, p\}} \alpha_i \alpha_j s_{X_{l+1-i}^*}(u, \delta) s_{X_{l+1-j}^*}(u, \delta) \right) \mu(du) d\delta
$$

$$
= \frac{1}{N} \int_0^1 \int_{S^0} \sum_{l=1}^{N+p-1} \left(\sum_{i=\max\{1, l+1-N\}}^{\min\{l, p\}} \alpha_i s_{X_{l+1-i}^*}(u, \delta) \right)^2 \mu(du) d\delta \geq 0
$$

Therefore $\widehat{\Gamma}$ is positive semi-definite.   □

For real time series $(X_t)_{t \in T}$ the matrix $\widehat{\Gamma}$ is different to the estimates from the classical Yule–Walker equations. From $\mathbb{E} X_t = 0$ it follows $\mathbb{E} X_t X_s = \mathbb{E} X_t X_s - \mathbb{E} X_t \mathbb{E} X_s = \text{Cov}(X_t, X_s)$ for a time series with real observations. For classical time series the estimations of the covariance can be used as elements of the estimated matrix. The values in $\widehat{\Gamma}$ for real valued time series are

$$
\widehat{\eta}(h) = \frac{1}{N} \sum_{i=1}^{N-h} \langle X_i, X_{i+h} \rangle = \frac{1}{N} \sum_{i=1}^{N-h} X_i X_{i+h},
$$

i.e. the estimates of products of random variables.

For fuzzy time series by the fuzziness of the mean the conditions $\mathbb{E} X_t^* = 0$ and $\langle \mathbb{E} X_t^*, \mathbb{E} X_s^* \rangle = 0$ are not possible.

The restriction to positive coefficients is not satisfying. A solution to this is the following: For all binary numbers with p digits $(e_1, \ldots, e_p) \in \{0, 1\}^p$ the solutions of

$$\mathbb{E}\, \rho_2^2 \left(X_t^*, \bigoplus_{i=1}^{p} \alpha_i \cdot \left((-1)^{e_i} \cdot X_{t-i}^* \right) \right) \to \min$$

for positive coefficients α_i, $i = 1\,(1)\,p$, are calculatd by quadratic optimization under side conditions. The symbol e_i represents the sign of α_i. For $e_i = 1$ and $\alpha_i \geq 0$ the ith component is $\alpha_i \cdot \left((-1)^{e_i} \cdot X_{t-i}^* \right) = \alpha_i \cdot \left(-X_{t-i}^* \right) = (-\alpha_i) \cdot X_{t-i}^*$, which corresponds to the muliplication of X_{t-i}^* with a negative coefficient. From the 2^p solutions for all possible combinations of the e_i, $i = 1\,(1)\,p$, the solution with minimal expected quadratic distance is taken. Let $\widehat{\alpha}_1, \ldots, \widehat{\alpha}_p$ be the coefficients of this best approximation, and $(e_1, \ldots, e_p)$ the corresponding vector of signs, then the vector $\left((-1)^{e_1} \widehat{\alpha}_1, \ldots, (-1)^{e_p} \widehat{\alpha}_p \right)$ contains the solution of the minimization problem (26.2).

The disadvantage of this method is the exponentially growing number of necessary calculations in p, so it is practicable only for small p.

Besides conditions (S1)–(S3) concerning the stochastic process another condition (S4) (after Remark 26.2) is necessary. The reason for that is explained below.

The linear model (26.1) is only suitable for small values of the parameter p. Another model which allows positive and negative coefficients from the beginning is the following:

Instead of allowing only positive or only negative coefficients α_i for $\alpha_i \cdot X_{t-i}^*$, $i = 1\,(1)\,p$, both kind of coefficients are allowed: $\alpha_i \cdot X_{t-i}^*$ with $\alpha_i \geq 0$ and $\beta_i \cdot X_{t-i}^*$ with $\beta_i \leq 0$, and considered simultaneously. The last expression can be written as $\beta_i \cdot (-X_{t-i}^*)$ with $\beta_i \geq 0$ for which in the following we write $-X_{t-i}^* = (-1) \cdot X_{t-i}^*$. Together for the approximation of X_t^* by the last p observations $X_{t-1}^*, \ldots, X_{t-p}^*$ the following form is used:

$$X_t^* \approx \bigoplus_{i=1}^{p} \alpha_i \cdot X_{t-i}^* \oplus \bigoplus_{i=1}^{p} \beta_i \cdot (-X_{t-i}^*) \qquad (26.6)$$

with $\alpha_i, \beta_i \geq 0$, $i = 1\,(1)\,p$. The condition for the determination of the coefficients is

$$\min_{(\alpha_1,\ldots,\alpha_p,\beta_1,\ldots,\beta_p)\in(\mathbb{R}_0^+)^{2p}} \mathbb{E}\, \rho_2^2 \left(X_t^*, \bigoplus_{i=1}^{p} \alpha_i \cdot X_{t-i}^* \oplus \bigoplus_{i=1}^{p} \beta_i \cdot (-X_{t-i}^*) \right) \qquad (26.7)$$

The approximation in this approach is better, i.e. the minimal expected quadratic distance is not greater than for the earlier approach (26.1). The corresponding coefficients $\left((-1)^{e_1} \widehat{\alpha}_1, \ldots, (-1)^{e_p} \widehat{\alpha}_p \right)$ of the optimal approximation in (26.1) for the approach (26.6) are

$$\alpha_i = \begin{cases} \widehat{\alpha}_i & \text{for} \quad e_i = 0 \\ 0 & \text{for} \quad e_i = 1 \end{cases} \quad \text{and} \quad \beta_i = \begin{cases} 0 & \text{for} \quad e_i = 0 \\ \widehat{\alpha}_i & \text{for} \quad e_i = 1. \end{cases}$$

For the special case of real valued time series $(X_t)_{t \in T}$ the models (26.1) and (26.6) are equivalent. This is seen from the following consideration: The minimal expected quadratic distance for model (26.6) is less than that for model (26.1). Let $(\widetilde{\alpha}_1, \ldots, \widetilde{\alpha}_p, \widetilde{\beta}_1, \ldots, \widetilde{\beta}_p) \in (\mathbb{R}_0^+)^{2p}$ be the coefficients of the optimal approximation for (26.6). Then the corresponding coefficients in model (26.1) are $\alpha_i = \widetilde{\alpha}_i - \widetilde{\beta}_i$, $i = 1\,(1)\,p$.

By the assumption of positive coefficients and Equation (23.10) the condition (26.7) is equivalent to

$$
\mathbb{E}\, \rho_2^2 \left(X_t^*, \bigoplus_{i=1}^{p} \alpha_i \cdot X_{t-i}^* \oplus \bigoplus_{i=1}^{p} \beta_i \cdot (-X_{t-i}^*) \right)
$$

$$
= \mathbb{E} \left(\langle X_t^*, X_t^* \rangle - 2 \left\langle X_t^*, \bigoplus_{i=1}^{p} \alpha_i \cdot X_{t-i}^* \oplus \bigoplus_{i=1}^{p} \beta_i \cdot (-X_{t-i}^*) \right\rangle \right.
$$

$$
\left. + \left\langle \bigoplus_{i=1}^{p} \alpha_i \cdot X_{t-i}^* \oplus \bigoplus_{i=1}^{p} \beta_i \cdot (-X_{t-i}^*), \bigoplus_{i=1}^{p} \alpha_i \cdot X_{t-i}^* \oplus \bigoplus_{i=1}^{p} \beta_i \cdot (-X_{t-i}^*) \right\rangle \right)
$$

$$
= \mathbb{E} \langle X_t^*, X_t^* \rangle - 2 \sum_{i=1}^{p} \alpha_i\, \mathbb{E} \langle X_t^*, X_{t-i}^* \rangle - 2 \sum_{i=1}^{p} \beta_i\, \mathbb{E} \langle X_t^*, -X_{t-i}^* \rangle
$$

$$
+ \sum_{i=1}^{p} \sum_{j=1}^{p} \alpha_i\, \alpha_j\, \mathbb{E} \langle X_{t-i}^*, X_{t-j}^* \rangle + 2 \sum_{i=1}^{p} \sum_{j=1}^{p} \alpha_i\, \beta_j\, \mathbb{E} \langle X_{t-i}^*, -X_{t-j}^* \rangle
$$

$$
+ \sum_{i=1}^{p} \sum_{j=1}^{p} \beta_i\, \beta_j\, \mathbb{E} \langle -X_{t-i}^*, -X_{t-j}^* \rangle \to \min.
$$

Using the notations

$$
\boldsymbol{\alpha}_2 = (\alpha_1, \ldots, \alpha_p, \beta_1, \ldots, \beta_p)
$$

$$
\boldsymbol{\gamma}_2 = \left(\mathbb{E} \langle X_t^*, X_{t-1}^* \rangle, \ldots, \mathbb{E} \langle X_t^*, X_{t-p}^* \rangle, \mathbb{E} \langle X_t^*, -X_{t-1}^* \rangle, \ldots, \mathbb{E} \langle X_t^*, -X_{t-p}^* \rangle \right)
$$

and the symmetric matrix

$$
\Gamma_2 = \begin{pmatrix}
\mathbb{E}\langle X_{t-1}^*, X_{t-1}^* \rangle & \cdots & \mathbb{E}\langle X_{t-1}^*, X_{t-p}^* \rangle & \mathbb{E}\langle X_{t-1}^*, -X_{t-1}^* \rangle & \cdots & \mathbb{E}\langle X_{t-1}^*, -X_{t-p}^* \rangle \\
\mathbb{E}\langle X_{t-2}^*, X_{t-1}^* \rangle & \cdots & \mathbb{E}\langle X_{t-2}^*, X_{t-p}^* \rangle & \mathbb{E}\langle X_{t-2}^*, -X_{t-1}^* \rangle & \cdots & \mathbb{E}\langle X_{t-2}^*, -X_{t-p}^* \rangle \\
\vdots & \ddots & \vdots & \vdots & \ddots & \vdots \\
\mathbb{E}\langle X_{t-p}^*, X_{t-1}^* \rangle & \cdots & \mathbb{E}\langle X_{t-p}^*, X_{t-p}^* \rangle & \mathbb{E}\langle X_{t-p}^*, -X_{t-1}^* \rangle & \cdots & \mathbb{E}\langle X_{t-p}^*, -X_{t-p}^* \rangle \\
\mathbb{E}\langle X_{t-1}^*, -X_{t-1}^* \rangle & \cdots & \mathbb{E}\langle X_{t-1}^*, -X_{t-p}^* \rangle & \mathbb{E}\langle -X_{t-1}^*, -X_{t-1}^* \rangle & \cdots & \mathbb{E}\langle -X_{t-1}^*, -X_{t-p}^* \rangle \\
\mathbb{E}\langle X_{t-2}^*, -X_{t-1}^* \rangle & \cdots & \mathbb{E}\langle X_{t-2}^*, -X_{t-p}^* \rangle & \mathbb{E}\langle -X_{t-2}^*, -X_{t-1}^* \rangle & \cdots & \mathbb{E}\langle -X_{t-2}^*, -X_{t-p}^* \rangle \\
\vdots & \ddots & \vdots & \vdots & \ddots & \vdots \\
\mathbb{E}\langle X_{t-p}^*, -X_{t-1}^* \rangle & \cdots & \mathbb{E}\langle X_{t-p}^*, -X_{t-p}^* \rangle & \mathbb{E}\langle -X_{t-p}^*, -X_{t-1}^* \rangle & \cdots & \mathbb{E}\langle -X_{t-p}^*, -X_{t-p}^* \rangle
\end{pmatrix}
$$

the above problem can be seen as the minimization of the quadratic form

$$
\mathbb{E}\,\rho_2^2\left(X_t^*, \bigoplus_{i=1}^{p}\alpha_i\cdot X_{t-i}^* \oplus \bigoplus_{i=1}^{p}\beta_i\cdot(-X_{t-i}^*)\right) = \alpha_2\,\Gamma_2\,\alpha_2^T - 2\,\gamma_2\,\alpha_2^T \to \min
$$

with side conditions $\alpha_i, \beta_i \geq 0,\ i = 1\,(1)\,p$.

Remark 26.2: The matrix Γ_2 is positive semi-definite.

Proof: For an arbitrary vector $\alpha = (\alpha_1, \ldots, \alpha_p, \beta_1, \ldots, \beta_p) \in \mathbb{R}^{2p}$, equivalent to the calculations in the proof of Remark 26.1 we have

$$
\alpha\,\Gamma_2\,\alpha^T = \mathbb{E}\int_0^1\int_{S^0}\left(\sum_{i=1}^{p}\alpha_i s_{X_{t-i}^*}(u,\delta) + \sum_{i=1}^{p}\beta_i s_{(-X_{t-i}^*)}(u,\delta)\right)^2 \mu(du)d\delta \geq 0.
$$

For applications the conditions (S1)–(S3) are not sufficient for the model (26.6). For the covariance of two fuzzy random quantities X_t^* and X_s^* it is possible that $\mathrm{Cov}\,(X_t^*, -X_s^*) \neq -\mathrm{Cov}\,(X_t^*, X_s^*)$, (cf. Proposition 25.1). By condition (S3) it is not provided that $\mathrm{Cov}\,(X_t^*, -X_s^*)$ depends only on the time difference $t - s$. Therefore it is assumed that the basic fuzzy stochastic process $(X_t^*)_{t\in T}$ is fulfilling the following condition:

(S4) $\mathrm{Cov}\left(X_t^*, -X_s^*\right) = \gamma_-\,(t - s)$

By the equality

$$
\begin{aligned}
\mathrm{Cov}\,(X_t^*, -X_s^*) &= \mathbb{E}\,\langle X_t^*, -X_s^*\rangle + \langle\mathbb{E}\,X_t^*, \mathbb{E}\,(-X_s^*)\rangle \\
&= \mathbb{E}\,\langle X_t^*, -X_s^*\rangle + \langle\mathbb{E}\,X_t^*, -\mathbb{E}\,X_s^*\rangle
\end{aligned}
$$

also the scalar product

$$
\begin{aligned}
\eta_-(t - s) &= \mathbb{E}\,\langle X_t^*, -X_s^*\rangle = \mathrm{Cov}\,(X_t^*, -X_s^*) - \langle\mathbb{E}\,X_t^*, -\mathbb{E}\,X_s^*\rangle \\
&= \gamma_-(t - s) - \langle\mu^*, -\mu^*\rangle
\end{aligned}
$$

depends only on the difference $t - s$. An estimator for the quantity $\eta_-\,(t - s)$, analogously to the estimation of $\eta\,(t - s)$, using the notation $h = |t - s|$ is

$$
\widehat{\eta}_-\,(h) = \frac{1}{N}\sum_{i=1}^{N-h}\langle X_i^*, -X_{i+h}^*\rangle = \frac{1}{N}\sum_{i=\max\{1,1-(t-s)\}}^{\min\{N,N-(t-s)\}}\langle X_i^*, -X_{i+(t-s)}^*\rangle. \tag{26.8}
$$

Lemma 26.2: The following matrix based on the estimators (26.5) and (26.8)

$$
\widehat{\Gamma}_2 =
\begin{pmatrix}
\widehat{\eta}(0) & \cdots & \widehat{\eta}(p-1) & \widehat{\eta}_-(0) & \cdots & \widehat{\eta}_-(p-1) \\
\widehat{\eta}(1) & \cdots & \widehat{\eta}(p-2) & \widehat{\eta}_-(1) & \cdots & \widehat{\eta}_-(p-2) \\
\vdots & \ddots & \vdots & \vdots & \ddots & \vdots \\
\widehat{\eta}(p-1) & \cdots & \widehat{\eta}(0) & \widehat{\eta}_-(p-1) & \cdots & \widehat{\eta}_-(0) \\
\widehat{\eta}_-(0) & \cdots & \widehat{\eta}_-(p-1) & \widehat{\eta}(0) & \cdots & \widehat{\eta}(p-1) \\
\widehat{\eta}_-(1) & \cdots & \widehat{\eta}_-(p-2) & \widehat{\eta}(1) & \cdots & \widehat{\eta}(p-2) \\
\vdots & \ddots & \vdots & \vdots & \ddots & \vdots \\
\widehat{\eta}_-(p-1) & \cdots & \widehat{\eta}_-(1) & \widehat{\eta}(p-1) & \cdots & \widehat{\eta}(0)
\end{pmatrix}
$$

is positive semi-definite.

Proof: Analogous to the calculations in the proof of Remark 26.1, for arbitrary vector $\alpha = (\alpha_1, \ldots, \alpha_p, \beta_1, \ldots, \beta_p) \in \mathbb{R}^{2p}$ we have

$$
\alpha \widehat{\Gamma}_2 \alpha^T = \frac{1}{N} \int_0^1 \int_{S^0} \sum_{l=1}^{N+p-1}
$$

$$
\times \left(\sum_{i=\max\{1,l+1-N\}}^{\min\{l,p\}} \left(\alpha_i \, s_{X^*_{l-i}}(u,\delta) + \beta_i \, s_{(-X^*_{l-i})}(u,\delta) \right) \right)^2 \mu(d\,u)\,d\,\delta \geq 0.
$$

□

The assumptions (26.1) and (26.6) have the following disadvantages: The fuzziness of the sum of two fuzzy numbers is essentially the sum of the fuzziness of the first fuzzy number and the second fuzzy number. For increasing coefficients α_i, $i = 1\,(1)\,p$, in model (26.1), or α_i and β_i, $i = 1\,(1)\,p$, in model (26.6), the fuzziness of the sum increases.

By the definition of the metric $\rho_2(\cdot, \cdot)$ and (23.8) follows that the distance $\rho_2(x^*, y^*)$ between two fuzzy numbers $x^*, y^* \in \mathcal{F}_2(\mathbb{R})$ depends on two factors: by the position described by the Steiner points σ_{x^*} and σ_{y^*}, and by the fuzziness of x^* and y^*. In both models (26.1) and (26.6) the positive coefficients have to be found which minimize the difference in position as well as the difference in the fuzziness between the observations X^*_t and the formed sum. By condition (S1) we obtain

$$
\sum_{i=1}^p \alpha_i \approx 1 \quad \text{and} \quad \sum_{i=1}^p (\alpha_i + \beta_i) \approx 1.
$$

The limitation of the coefficients implies a limitation for the approximation of the position.

In the case of real valued time series and using the Yule–Walker equations, the coefficients in (26.3) have no restrictions. The absolute values of the values $(\widehat{\alpha}_1, \ldots, \widehat{\alpha}_p) \in \mathbb{R}^p$ can be greater than 1.

A third model, which allows a more flexible approximation of the location, is

$$X_t^* \oplus \bigoplus_{i=1}^p \alpha_i \cdot X_{t-i}^* \approx \bigoplus_{i=1}^p \beta_i \cdot X_{t-i}^* \tag{26.9}$$

with $\alpha_i, \beta_i \geq 0, i = 1\,(1)\,p$. The idea for this model is the following: In the case of real valued observations the calculated value obtained from (26.3),

$$X_t \approx \sum_{i=1}^p \widehat{\alpha}_i \, X_{t-i}$$

with $\alpha_i \in \mathbb{R}, i = 1\,(1)\,p$, can be written in the following way:

$$X_t + \sum_{i=1}^p \widehat{\alpha}_i^- \, X_{t-i} \approx \sum_{i=1}^p \widehat{\alpha}_i^+ \, X_{t-i}$$

with $\widehat{\alpha}_i^+ = \max\{\widehat{\alpha}_i, 0\}$ and $\widehat{\alpha}_i^- = \max\{-\widehat{\alpha}_i, 0\}$.

The condition for the determination of the coefficients in (26.9) is the following:

$$\min_{(\alpha_1,\ldots,\alpha_p,\beta_1,\ldots,\beta_p)\in(\mathbb{R}_0^+)^{2p}} \mathbb{E}\,\rho_2^2 \left(X_t^* \oplus \bigoplus_{i=1}^p \alpha_i \cdot X_{t-i}^*, \bigoplus_{i=1}^p \beta_i \cdot X_{t-i}^* \right) \tag{26.10}$$

This minimization can be written in the following way:

$$\mathbb{E}\,\rho_2^2 \left(X_t^* \oplus \bigoplus_{i=1}^p \alpha_i \cdot X_{t-i}^*, \bigoplus_{i=1}^p \beta_i \cdot X_{t-i}^* \right)$$

$$= \mathbb{E}\left(\left\langle X_t^* \oplus \bigoplus_{i=1}^p \alpha_i \cdot X_{t-i}^*, X_t^* \oplus \bigoplus_{i=1}^p \alpha_i \cdot X_{t-i}^* \right\rangle \right.$$

$$\left. -2\left\langle X_t^* \oplus \bigoplus_{i=1}^p \alpha_i \cdot X_{t-i}^*, \bigoplus_{i=1}^p \beta_i \cdot X_{t-i}^* \right\rangle + \left\langle \bigoplus_{i=1}^p \beta_i \cdot X_{t-i}^*, \bigoplus_{i=1}^p \beta_i \cdot X_{t-i}^* \right\rangle \right)$$

$$= \mathbb{E}\,\langle X_t^*, X_t^* \rangle + 2\sum_{i=1}^p \alpha_i\,\mathbb{E}\,\langle X_t^*, X_{t-i}^* \rangle + \sum_{i=1}^p \sum_{j=1}^p \alpha_i\,\alpha_j\,\mathbb{E}\,\langle X_{t-i}^*, X_{t-j}^* \rangle$$

$$-2\sum_{i=1}^p \beta_i\,\mathbb{E}\,\langle X_t^*, X_{t-i}^* \rangle - 2\sum_{i=1}^p \sum_{j=1}^p \alpha_i\,\beta_j\,\mathbb{E}\,\langle X_{t-i}^*, X_{t-j}^* \rangle$$

$$+ \sum_{i=1}^p \sum_{j=1}^p \beta_i\,\beta_j\,\mathbb{E}\,\langle X_{t-i}^*, X_{t-j}^* \rangle \to \min$$

Using the vectors

$$\boldsymbol{\alpha}_3 = (\alpha_1, \ldots, \alpha_p, \beta_1, \ldots, \beta_p)$$
$$\boldsymbol{\gamma}_3 = \left(\mathbb{E}\, \langle X_t^*, X_{t-1}^* \rangle, \ldots, \mathbb{E}\, \langle X_t^*, X_{t-p}^* \rangle, -\mathbb{E}\, \langle X_t^*, X_{t-1}^* \rangle, \ldots, -\mathbb{E}\, \langle X_t^*, X_{t-p}^* \rangle \right)$$

and the symmetric matrix

$$\Gamma_3 = \begin{pmatrix}
\mathbb{E}\, \langle X_{t-1}^*, X_{t-1}^* \rangle & \cdots & \mathbb{E}\, \langle X_{t-1}^*, X_{t-p}^* \rangle & -\mathbb{E}\, \langle X_{t-1}^*, X_{t-1}^* \rangle & \cdots & -\mathbb{E}\, \langle X_{t-1}^*, X_{t-p}^* \rangle \\
\mathbb{E}\, \langle X_{t-2}^*, X_{t-1}^* \rangle & \cdots & \mathbb{E}\, \langle X_{t-2}^*, X_{t-p}^* \rangle & -\mathbb{E}\, \langle X_{t-2}^*, X_{t-1}^* \rangle & \cdots & -\mathbb{E}\, \langle X_{t-2}^*, X_{t-p}^* \rangle \\
\vdots & \ddots & \vdots & \vdots & \ddots & \vdots \\
\mathbb{E}\, \langle X_{t-p}^*, X_{t-1}^* \rangle & \cdots & \mathbb{E}\, \langle X_{t-p}^*, X_{t-p}^* \rangle & -\mathbb{E}\, \langle X_{t-p}^*, X_{t-1}^* \rangle & \cdots & -\mathbb{E}\, \langle X_{t-p}^*, X_{t-p}^* \rangle \\
-\mathbb{E}\, \langle X_{t-1}^*, X_{t-1}^* \rangle & \cdots & -\mathbb{E}\, \langle X_{t-1}^*, X_{t-p}^* \rangle & \mathbb{E}\, \langle X_{t-1}^*, X_{t-1}^* \rangle & \cdots & \mathbb{E}\, \langle X_{t-1}^*, X_{t-p}^* \rangle \\
-\mathbb{E}\, \langle X_{t-2}^*, X_{t-1}^* \rangle & \cdots & -\mathbb{E}\, \langle X_{t-2}^*, X_{t-p}^* \rangle & \mathbb{E}\, \langle X_{t-2}^*, X_{t-1}^* \rangle & \cdots & \mathbb{E}\, \langle X_{t-2}^*, X_{t-p}^* \rangle \\
\vdots & \ddots & \vdots & \vdots & \ddots & \vdots \\
-\mathbb{E}\, \langle X_{t-p}^*, X_{t-1}^* \rangle & \cdots & -\mathbb{E}\, \langle X_{t-p}^*, X_{t-p}^* \rangle & \mathbb{E}\, \langle X_{t-p}^*, X_{t-1}^* \rangle & \cdots & \mathbb{E}\, \langle X_{t-p}^*, X_{t-p}^* \rangle
\end{pmatrix}$$

the minimization is a minimization of a quadratic form

$$\mathbb{E}\, \rho_2^2 \left(X_t^* \oplus \bigoplus_{i=1}^{p} \alpha_i \cdot X_{t-i}^*, \bigoplus_{i=1}^{p} \beta_i \cdot (-X_{t-i}^*) \right) = \boldsymbol{\alpha}_3 \, \Gamma_3 \, \boldsymbol{\alpha}_3^T + 2 \, \boldsymbol{\gamma}_3 \, \boldsymbol{\alpha}_3^T \rightarrow \min$$

with side conditions $\alpha_i, \beta_i \geq 0, i = 1\,(1)\,p$.

Remark 26.3: The matrix Γ_3 is positive semi-definite.

Proof: For an arbitrary vector $\boldsymbol{\alpha} = (\alpha_1, \ldots, \alpha_p, \beta_1, \ldots, \beta_p) \in \mathbb{R}^{2p}$ an equivalent transformation yields

$$\boldsymbol{\alpha}\, \Gamma_2\, \boldsymbol{\alpha}^T = \mathbb{E} \int_0^1 \int_{S^0} \left(\sum_{i=1}^{p} \alpha_i\, s_{X_{t-i}^*}(u, \delta) - \sum_{i=1}^{p} \beta_i\, s_{X_{t-i}^*}(u, \delta) \right)^2 \mu(d\,u)\, d\,\delta \geq 0.$$

$\square$

In order to estimate the matrix Γ_3 the biased estimator (26.5) is used. Similar to Proposition 26.1 it can be proved that the matrix $\widehat{\Gamma}_3$, formed by these estimators, is positive semi-definite.

Let $(\widehat{\alpha}_1, \ldots, \widehat{\alpha}_p, \widehat{\beta}_1, \ldots, \widehat{\beta}_p) \in (\mathbb{R}_0^+)^{2p}$ be the solution of (26.10), then the estimator $\widehat{X}_t^*$ is calculated by

$$\widehat{X}_t^* = \left(\bigoplus_{i=1}^{p} \widehat{\beta}_i \cdot X_{t-i}^* \right) \sim_H \left(\bigoplus_{i=1}^{p} \widehat{\alpha}_i \cdot X_{t-i}^* \right).$$

Remark 26.4: In the case of real valued observations the three models (26.1), (26.6) and (26.9) are equivalent.

As mentioned above, the model (26.9) allows a more flexible approximation of location and fuzziness. However, for a fuzzy time series $(x_t^*)_{t \in T}$ with a small number N and estimated parameters $(\widehat{\alpha}_1, \ldots, \widehat{\alpha}_p, \widehat{\beta}_1, \ldots, \widehat{\beta}_p)$ by the biased estimates of covariances the fuzziness $\widehat{\alpha}_1 \cdot X_{t-1}^* \oplus \cdots \oplus \widehat{\alpha}_p \cdot X_{t-p}^*$ can be greater than the fuzziness of $\widehat{\beta}_1 \cdot X_{t-1}^* \oplus \cdots \oplus \widehat{\beta}_p \cdot X_{t-p}^*$. In this situation the Hukuhara difference of these sums and therefore the estimate of X_t^* is a real number. This problem does not exist in the more flexible models (26.1) and (26.6).

An interesting question in connection with linear approximation is the choice of p. The larger p, the better the approximation. But with increasing p the computational complexity for the calculation of the coefficients increases. The most appropriate value is the value $\widehat{p}$ for which a larger value of p does not bring an essential improvement of the approximation. In order to determine $\widehat{p}$ procedures of model selection from classical time series analysis can be used [see Schlittgen and Streitberg (1984)]. A suitable quantity for the quality of an approximation is the least expected sum of squared distances.

For model (26.1) the method to determine p is

$$\min_{p \geq 1} \left\{ \min_{(\alpha_1, \ldots, \alpha_p) \in \mathbb{R}^p} \mathbb{E} \, \rho_2^2 \left(X_t^*, \bigoplus_{i=1}^p \alpha_i \cdot X_{t-i}^* \right) + p \right\} \, .$$

If an increase of p brings no essential improvement for the approximation then the first part of the sum stays constant.

With the determination of the most appropriate p, the question of how long linear influences from the past are detectable at time t is also answered.

The best linear prediction $\widehat{X}_{N+h}^*$ at time N for the time $N + h$, with $h > 0$, from the $p + 1$ observations $X_N^*, \ldots, X_{N-p}^*$ can be calculated for the above considered models. For model (26.1) for the best linear prediction X_{N+h}^* are obtained from

$$\min_{(\alpha_0, \ldots, \alpha_p) \in \mathbb{R}^{p+1}} \mathbb{E} \, \rho_2^2 \left(X_{N+h}^*, \bigoplus_{i=0}^p \alpha_i \cdot X_{N-i}^* \right) \, .$$

For model (26.6) by

$$\min_{(\alpha_0, \ldots, \alpha_p, \beta_0, \ldots, \beta_p) \in (\mathbb{R}_0^+)^{2(p+1)}} \mathbb{E} \, \rho_2^2 \left(X_{N+h}^*, \bigoplus_{i=0}^p \alpha_i \cdot X_{N-i}^* \oplus \bigoplus_{i=0}^p \beta_i \cdot (-X_{N-i}^*) \right) \, .$$

For model (26.9) by

$$\min_{(\alpha_0, \ldots, \alpha_p, \beta_0, \ldots, \beta_p) \in (\mathbb{R}_0^+)^{2(p+1)}} \mathbb{E} \, \rho_2^2 \left(X_{N+h}^* \oplus \bigoplus_{i=0}^p \alpha_i \cdot X_{N-i}^*, \bigoplus_{i=0}^p \beta_i \cdot X_{N-i}^* \right) \, .$$

26.2   Remarks concerning Kalman filtering

Many processes can be modeled by a multivariate state-space model [see Catlin (1989) and Grewal and Andrews (2001)]. This linear dynamic system consists of a system equation

$$x_{t+1} = A_t\, x_t + B_t\, \varepsilon_{t+1}, \qquad t \geq 0 \tag{26.11}$$

and an observation equation

$$y_t = C_t\, x_t + \eta_t. \tag{26.12}$$

The vector $x_t \in \mathbb{R}^n$ describes the state of the system at time t, and the system equation models the dynamics of the system. The dynamics is influenced by the disturbance term $\varepsilon_t \in \mathbb{R}^l$. In this model the state of the system cannot be observed directly but only a linear function $y_t \in \mathbb{R}^m$ can be observed. This linear dependence changes with time and is described by Equation (26.12) which is also influenced by a disturbance $\eta_t \in \mathbb{R}^m$.

$(\varepsilon_t)_{t \geq 0}$ and $(\eta_t)_{t \geq 0}$ are sequences of independent stochastic vectors with $\mathrm{Var}(\varepsilon_t) = Q_t \in \mathbb{R}^{l \times l}$ and $\mathrm{Var}(\eta_t) = R_t \in \mathbb{R}^{m \times m}$. $(A_t)_{t \geq 0}$, $(B_t)_{t \geq 0}$ and $(C_t)_{t \geq 0}$ with $A_t \in \mathbb{R}^{n \times n}$, $B_t \in \mathbb{R}^{n \times l}$ and $C_t \in \mathbb{R}^{m \times n}$ are sequences of known matrices.

The advantage of this model is that the estimate for x_t can be calculated recursively from the estimate for x_{t-1} and the last observation y_t. The estimate $\widehat{x}_t$ is the best linear approximation of x_t based on the observed values $y_1, \dots, y_t$. A measure for the estimation error and the prediction error is the variance–covariance matrix

$$P_t = \mathrm{Var}\,(x_t - \widehat{x}_t)\,.$$

The recursive calculation of the estimate $\widehat{x}_t$ from $\widehat{x}_{t-1}$ and y_t is in two steps and is called the *discrete Kalman filter*.

The first step is the so-called *prediction step*. Here the best estimate $\widetilde{x}_t$ for x_t based on $\widehat{x}_{t-1}$ is calculated:

(Ps1) $\widetilde{x}_t = A_{t-1} \widehat{x}_{t-1}$

(Ps2) $\widetilde{y}_t = C_t \widetilde{x}_t$

(Ps3) $\widetilde{P}_t = A_{t-1}\, P_{t-1}\, A_{t-1}^T + B_{t-1}\, Q_{t-1}\, B_{t-1}^T.$

After observation of y_t an improvement of the estimate $\widetilde{x}_t$ follows by the so-called *correction step*:

(Ks1) $K_t = \widetilde{P}_t\, C_t^T \left(C_t\, \widetilde{P}_t\, C_t^T + R_t \right)^{-1}$

(Ks2) $\widehat{x}_t = \widetilde{x}_t + K_t \left(y_t - \widetilde{y}_t \right)$

(Ks3) $P_t = \widetilde{P}_t - K_t\, C_t\, \widetilde{P}_t.$

A h-step prediction $\widetilde{\pmb{x}}_{t,h}$, $h \geq 1$, of the state at time t for time $t + h$ is obtained by

$$\widetilde{\pmb{x}}_{t,1} = A_t\,\widehat{\pmb{x}}_t = \widetilde{\pmb{x}}_{t+1} \qquad \text{and} \tag{26.13}$$

$$\widetilde{\pmb{x}}_{t,h} = A_{t+h-1}\,\widetilde{\pmb{x}}_{t,h-1}, \qquad h \geq 2. \tag{26.14}$$

An extension of the discrete Kalman filter to the situation of fuzzy observations $\pmb{y}_1^*, \ldots, \pmb{y}_t^*$ was given by Schnatter (1990). In this situation the state of the system is estimated by a fuzzy quantity $\widehat{\pmb{x}}_t^*$. Basic for the extension is the following composition of the equations (Ps1), (Ps2), (Ps3), (Ks1) and (Ks2):

$$\widehat{\pmb{x}}_t = \widetilde{\pmb{x}}_t + K_t\left(\pmb{y}_t - \widetilde{\pmb{y}}_t\right) \overset{\text{(Ps2)}}{=} \widetilde{\pmb{x}}_t + K_t\left(\pmb{y}_t - C_t\,\widetilde{\pmb{x}}_t\right)$$

$$\overset{\text{(Ps1)}}{=} A_{t-1}\,\widehat{\pmb{x}}_{t-1} + K_t\left(\pmb{y}_t - C_t\,A_{t-1}\,\widehat{\pmb{x}}_{t-1}\right)$$

$$\overset{\text{(Ks1)}}{=} A_{t-1}\,\widehat{\pmb{x}}_{t-1} + \widetilde{P}_t\,C_t^T\left(C_t\,\widetilde{P}_t\,C_t^T + R_t\right)^{-1}\left(\pmb{y}_t - C_t\,A_{t-1}\,\widehat{\pmb{x}}_{t-1}\right)$$

$$\overset{\text{(Ps3)}}{=} A_{t-1}\,\widehat{\pmb{x}}_{t-1} + \left(A_{t-1}\,P_{t-1}\,A_{t-1}^T + B_{t-1}\,Q_{t-1}\,B_{t-1}^T\right)C_t^T$$

$$\left(C_t\left(A_{t-1}\,P_{t-1}\,A_{t-1}^T + B_{t-1}\,Q_{t-1}\,B_{t-1}^T\right)C_t^T + R_t\right)^{-1}\left(\pmb{y}_t - C_t\,A_{t-1}\,\widehat{\pmb{x}}_{t-1}\right)$$

The estimate $\widehat{\pmb{x}}_t$ is a linear function of $\widehat{\pmb{x}}_{t-1}$ and $\pmb{y}_t$. The situation corresponds to a general linear vector-valued function $L: \mathbb{R}^{n_1} \times \cdots \times \mathbb{R}^{n_m} \to \mathbb{R}^n$ with

$$\pmb{y} = L\left(\pmb{x}_1, \ldots, \pmb{x}_m\right) = \sum_{i=1}^{m} A_i\,\pmb{x}_i \tag{26.15}$$

where $\pmb{y} \in \mathbb{R}^n$, $\pmb{x}_i \in \mathbb{R}^{n_i}$, $i = 1\,(1)\,m$, and $A_i \in \mathbb{R}^{n \times n_i}$, $i = 1\,(1)\,m$.

If the components of the vectors $\pmb{x}_i = (x_{i,1}, \ldots, x_{i,n_i})^T$ are fuzzy with $x_{i,j}^* \in \mathcal{F}(\mathbb{R})$, $i = 1\,(1)\,m$, $j = 1\,(1)\,n_i$, then the result $\pmb{y}^* = L(\pmb{x}_1^*, \ldots, \pmb{x}_m^*)$ is a fuzzy vector, and in order to calculate $\pmb{y}^*$ the extension principle is applied. There is a simple presentation of the result.

Theorem 26.1: Let $L(\cdot, \ldots, \cdot)$ be a vector-valued function of type (26.15) and fuzzy observations $\pmb{x}_i^*$ with components $(x_{i,j}^*)_{i=1\,(1)\,m, j=1\,(1)\,n_i}$ whose δ-cuts are $C_\delta(x_{i,j}^*) = [\underline{x}_{i,j,\delta}, \overline{x}_{i,j,\delta}]$, $\delta \in (0; 1]$, and let $\underline{C}_\delta(\pmb{x}_i^*) = (\underline{x}_{i,1,\delta}, \ldots, \underline{x}_{i,n_i,\delta})^T \in \mathbb{R}^{n_i}$ and $\overline{C}_\delta(\pmb{x}_i^*) = (\overline{x}_{i,1,\delta}, \ldots, \overline{x}_{i,n_i,\delta})^T \in \mathbb{R}^{n_i}$ be vectors whose components are the lower and upper limits of the δ-cuts of the observations. The vectors $\underline{C}_\delta(\pmb{y}^*) \in \mathbb{R}^m$ and $\overline{C}_\delta(\pmb{y}^*) \in \mathbb{R}^m$ whose components are the limits of the δ-cuts of $\pmb{y}^* = L(\pmb{x}_1^*, \ldots, \pmb{x}_m^*)$ with

$$\begin{aligned} \left(A_i^+\right)_{lk} &= \max\{(A_i)_{lk}, 0\} \\ \left(A_i^-\right)_{lk} &= \min\{(A_i)_{lk}, 0\} \end{aligned}, \qquad i = 1\,(1)\,m, \; l = 1\,(1)\,n, \; k = 1\,(1)\,n_i,$$

and

$$\underline{C_\delta}(y^*) = \sum_{i=1}^{m} A_i^+ \underline{C_\delta}(x_i^*) + \sum_{i=1}^{m} A_i^- \overline{C_\delta}(x_i^*) \qquad (26.16)$$

and

$$\overline{C_\delta}(y^*) = \sum_{i=1}^{m} A_i^- \underline{C_\delta}(x_i^*) + \sum_{i=1}^{m} A_i^+ \overline{C_\delta}(x_i^*). \qquad (26.17)$$

For the proof see Schnatter (1990).

The generalized discrete Kalman filter for fuzzy observations $y_1^*, \ldots, y_t^*$ calculates the fuzzy estimation $\hat{x}_t^*$ for the system state at time t from the fuzzy estimate $\hat{x}_{t-1}^*$ and the observation y_t^* based on Theorem 26.1 and the linear relationship between x_t, x_{t-1} and y_t.

A disadvantage of this generalization is the possible increase of fuzziness in the results. In particular, for matrices with large components the result for linear vector-valued functions (26.15) can have remarkable fuzziness as shown in the following example.

Example 26.1 For the matrices

$$A_1 = \begin{pmatrix} 5 & -4 \\ -3 & 4 \end{pmatrix}, \qquad A_2 = \begin{pmatrix} 2 & -3 \\ -7 & 7 \end{pmatrix}$$

and fuzzy vectors x_1^* and x_2^* whose components have characterizing functions $\xi_{x_{1,1}^*}(x) = I_{[0;0.1]}(x)$, $\xi_{x_{1,2}^*}(x) = I_{[-0.1;0]}(x)$, $\xi_{x_{2,1}^*}(x) = I_{[0;0.1]}(x)$ and $\xi_{x_{2,2}^*}(x) = I_{[0;0.2]}(x)$ the components of $y^* = L(x_1^*, x_2^*)$ calculated by (26.16) and (26.17) have the characterizing functions $\xi_{y_1^*}(x) = I_{[-0.6;1.1]}(x)$ and $\xi_{y_2^*}(x) = I_{[-0.7;2.1]}(x)$. Here the relative large elements in the matrices A_1 and A_2 cause the increase of fuzziness of the results relative to the fuzziness of the observations.

Remark 26.5: An alternative approach could be the Bayesian interpretation of the Kalman filter [see Meinhold and Singpurwalla (1983)]. This approach could be based on the generalized Bayes' theorem from Chapter 15. This is a topic of future research.

Part VIII

APPENDICES

The following four appendices A1 to A4 should be helpful for the reader in order to find information in an easy way.

Appendix A1 is a listing of symbols and abbreviations used in the book. As far as possible they are listed in alphabetical order. Greek letters in the order of phonetical translation to English. Mathematical symbols are listed at the end.

Appendix A2 consists of solutions to the problems given in the different chapters. If needed further details can be obtained from the author.

Appendix A3 is a glossary of the most important terms used throughout the book.

In Appendix A4 related literature is presented which is not referenced in the text. It gives also hints to approaches to the analysis of fuzzy data which are different to the approach in this book.

Part VIII

APPENDICES

A1

List of symbols and abbreviations

A	set
$\mathcal{A}$	event system
$[a; b]$	closed interval
$(a; b]$	left open and right closed interval
$a_{\delta,j}$	lower endpoint of the jth part of the δ-cut of a fuzzy number
$\{a_1, a_2, \dots\}$	set of elements $a_1, a_2, \dots$
B	Borel set
$\mathcal{B}$	system of Borel sets
$\mathcal{B}(\mathbb{R})$	one-dimensional Borel sets
$b_{\delta,j}$	upper endpoint of the jth part of the δ-cut of a fuzzy number
$Be(\alpha,\beta)$	beta-distribution with parameters α and β
C	real constant
$C_{0+}(y^*)$	support of y^*
$C_\delta[\xi(\cdot)]$	δ-cut of the characterizing function $\xi(\cdot)$
$C_\delta(\underline{x}^*)$	δ-cut of the fuzzy vector $\underline{x}^*$
$C_\delta(x^*)$	δ-cut of the fuzzy number x^*
$C_\delta[\xi(.,\dots,.)]$	δ-cut of the vector-characterizing function $\xi(.,\dots,.)$
d	decision
d_B	Bayesian decision
D^*	fuzzy data
$\mathcal{D}$	set of possible decisions
$\mathbb{E}_P L(\tilde{\theta}, d)$	expected loss of the decision d

Statistical Methods for Fuzzy Data   Reinhard Viertl
© 2011 John Wiley & Sons, Ltd

$\mathbb{E}_P U(\tilde{\theta}, d)$	expected utility
$\mathbb{E}_{P^*} L^*(\tilde{\theta}, d)$	fuzzy expected loss
$\mathbb{E}_{P^*} U^*(\tilde{\theta}, d)$	fuzzy expected utility
$\overline{\mathbb{E}}_\delta U^*(\tilde{\theta}, d)$	upper end of the δ-cut of a fuzzy expected utility
$\underline{\mathbb{E}}_\delta U^*(\tilde{\theta}, d)$	lower end of the δ-cut of a fuzzy expected utility
Ex_θ	exponential distribution
$\exp(\cdot)$	exponential function
$\overline{f}_\delta(\cdot)$	upper δ-level function of a fuzzy valued function
$\underline{f}_\delta(\cdot)$	lower δ-level function of a fuzzy valued function
$f(\cdot\|\theta)$	parametric stochastic model
$f(\cdot\|D)$	predictive density
$f(\cdot\|\theta), \theta \in \Theta$	continuous parametric stochastic model
$f(C_\delta(x^*))$	set of values of the function $f(\cdot)$
$f^*(\cdot\|D^*)$	fuzzy predictive density
$f^*(\cdot)$	fuzzy real valued function
$\mathcal{F}(\mathbb{R}^n)$	system of all fuzzy subsets of $\mathbb{R}^n$
$\mathcal{F}(\mathbb{R})$	system of all fuzzy subsets of $\mathbb{R}$
$\mathcal{F}_I([0;\infty))$	set of fuzzy intervals with non-negative support
$\mathcal{F}_I(\mathbb{R})$	set of all fuzzy intervals
$\mathcal{F}_N(\mathbb{R})$	normalized fuzzy subsets of $\mathbb{R}$
$\hat{F}_n(\cdot)$	empirical distribution function
$\hat{F}_n^*(\cdot)$	fuzzy empirical distribution function
$\hat{F}_n^{sm}(\cdot)$	smoothed empirical distribution function
$\hat{F}_{\delta,L}(\cdot)$	lower δ-level curve of $\hat{F}_n^*(\cdot)$
$\hat{F}_{\delta,U}(\cdot)$	upper δ-level curve of $\hat{F}_n^*(\cdot)$
$\overline{f}_{0+}(x)$	upper boundary of the support of $f^*(x)$
$\underline{f}_{0+}(x)$	lower boundary of the support of $f^*(x)$
$Gam(\alpha,\beta)$	gamma distribution with parameters α and β
$\overline{g}_\delta(\cdot)$	upper δ-level function of a fuzzy valued function
$\underline{g}_\delta(\cdot)$	lower δ-level function of a fuzzy valued function
$g^*(\cdot)$	fuzzy valued function
$\overline{h}_{n,\delta}(K_j)$	upper boundary of the δ-cut of $h_{n^*}(K_j)$
$\underline{h}_{n,\delta}(K_j)$	lower boundary of the δ-cut of $h_n^*(K_j)$
$h'(\cdot)$	derivative of the function $h(\cdot)$
$h_n^*(K_j)$	fuzzy relative frequency of the class K_j
$h_n^*(B, \omega)$	fuzzy relative frequency of B
$h_n(A)$	relative frequency of an event A
$h_n(K_j)$	relative frequency of the class K_j
$\mathcal{H}$	hypothesis
HPD	highest a posteriori density
inf	infimum
$I_A(\cdot)$	indicator function of the set A
$[\underline{I}_\delta; \overline{I}_\delta]$	δ-cut of $\mathcal{J}^*$
$\mathcal{J}^*$	integral of a fuzzy function

$\kappa(\ldots)$	confidence function
$\overline{\kappa}(\ldots)$	upper end of a confidence interval
$\underline{\kappa}(\ldots)$	lower end of a confidence interval
$L(.,.)$	loss function
$L^{-1}(\cdot)$	pseudo inverse of the real function $L(\cdot)$
$\overline{l}_\delta(\cdot\,;\underline{x}^*)$	upper δ-level function of the likelihood function
$\underline{l}_\delta(\cdot\,;\underline{x}^*)$	lower δ-level function of the likelihood function
$l(\cdot\,; x_1\ldots, x_n)$	likelihood function
ln	logarithmus naturalis (basis e)
$(M, \mathcal{A})$	measurable space
$(M, \mathcal{A}, \mu)$	measure space
m_k	kth sample moment
M_X^n	sample space of X
M_X	observation space of X
$M \setminus A$	set difference
$M_X^n = M_X \times M_X \times \ldots \times M_X$	Cartesian product of n copies of M_X, sample space
$N(\mu, \sigma^2)$	normal distribution with parameters μ and σ^2
μ_o	hypothetical mean
N^c	complement of the set N
$(N, \mathcal{Q})$	measurable space
N	set
Pr $(A\|B)$	conditional probability
Pr	probability
$\mathcal{P}(\Theta)$	power set of Θ
$\mathbb{P}$	set of possible probability distributions
$p(\cdot\|D)$	predictive probability based on data D
$p(\cdot\|\theta), \theta \in \Theta$	discrete parametric stochastic model
$p^*(\cdot\|D)$	fuzzy predictive probability based on fuzzy data D^*
p^*	fuzzy probability
P	probability distribution
P^*	fuzzy probability distribution
$P_\theta, \theta \in \Theta$	parametric family of probability distributions
P_o	true probability distribution
$\overline{P}_\delta(B)$	upper end of the δ-cut of $P^*(B)$
$\underline{P}_\delta(B)$	lower end of the δ-cut of $P^*(B)$
$\overline{p}_\delta$	upper end of the δ-cut of a fuzzy probability p^*
$\underline{p}_\delta$	lower end of the δ-cuts of a fuzzy probability p^*
$\{P_\theta : \theta \in \Theta\}$	parametric family of probability distributions
$(P) \int f^*(x)\,\mathrm{d}x$	probability integral
$\mathbb{Q}$	set of rational numbers
q_p^*	fuzzy p-fractile
$\mathbb{R}$	set of real numbers
$\mathbb{R}^k$	k-dimensional Euclidean space
r	empirical correlation coefficient

r^*	fuzzy empirical correlation coefficient
$R^{-1}(\cdot)$	pseudo inverse of the real function $R(\cdot)$
$(\mathbb{R}, \mathcal{B}, \lambda)$	Lebesgue measure space
$s(X_1, \ldots, X_n)$	statistic of the sample $X_1, \ldots, X_n$
$\text{supp}(f(\cdot))$	support of the function $f(\cdot)$
sup	supremum
$S_n^{ad}(\cdot)$	adapted cumulative sum
S_δ	set of classical probability densities
S_n	sample dispersion
S_n^*	fuzzy sample dispersion
S_n^2	sample variance
$T(.,.)$	t-norm, triangular norm
$t^*(m,s,l,r)$	trapezoidal fuzzy number
t^*	fuzzy value of a test statistic
t_{crit}	critical value for a test statistic
$U(\cdot,\cdot)$	utility function
$U^*(\cdot,\cdot)$	fuzzy utility function
$\overline{U}_\delta(.,.)$	upper δ-level curve of a fuzzy utility function
$\underline{U}_\delta(.,.)$	lower δ-level curve of a fuzzy utility function
x	real number
x^*	fuzzy number
$x_{\min}^*$	minimum of fuzzy numbers
$x_{\max}^*$	maximum of fuzzy numbers
X	stochastic quantity
$X \sim P_\theta, \theta \in \Theta$	parametric stochastic model
$X\|D$	predictive distribution of X given data D
X^*	fuzzy random variable
$X_\delta(\omega)$	δ-cut of a fuzzy random variable
$[\underline{x}; \overline{x}]$	closed interval
$\overline{X}_\delta(\omega)$	upper end of a fuzzy random interval
$\underline{X}_\delta(\omega)$	lower end of a fuzzy random interval
$\overline{x}_n$	sample mean
$\overline{x}_{\delta,i}$	upper end of the ith part of the δ-cut of a fuzzy number
$\underline{x}_{\delta,i}$	lower end of the ith part of the δ-cut of a fuzzy number
$\underline{x}^*$	fuzzy sector
$\underline{x} = (x_1, \ldots, x_n)$	real vector
$(X_n^*)_{n \in N}$	sequence of fuzzy random variables
$(x_1^*, \ldots, x_n^*)$	vector of fuzzy numbers
$\Gamma(\cdot)$	gamma function
Θ	parameter space
$\Theta \times \mathcal{D}$	Cartesian product of Θ and $\mathcal{D}$
$\Theta_{H,1-\alpha}$	HPD-region for the parameter θ
$\Theta_{1-\alpha}$	confidence region for θ with confidence level $1-\alpha$

$\Theta_{1-\alpha}^*$	fuzzy Bayesian confidence region
$\Theta_{\underline{x},1-\alpha}$	Bayesian confidence region for θ based on the classical sample $\underline{x}$
$\displaystyle\bigcup_{i=1}^{n}$	union operation for n sets
$\chi(\cdot)$	characterizing function
$\chi_{2n;p}^2$	p-fractile of the chi-square distribution with $2n$ degrees of freedom
χ_{2n}^2	chi-square distribution with $2n$ degrees of freedom
$\varnothing$	empty set
$\eta_j(\cdot)$	characterizing function
$\exists$	existence operator
$\forall$	for all' operator
$\hat{\theta}$	estimator for θ
$\hat{\theta}^*$	fuzzy estimator for θ
$\displaystyle\not\!\int$	fuzzy valued integral
$\displaystyle\not\!\int g^*(x)\,\mathrm{d}x$	integral of a fuzzy valued function
$\displaystyle\bigoplus_{i=1}^{n}$	sum operator for fuzzy numbers
$\mu(\cdot)$	measure
$\overline{\pi}_\delta(\cdot)$	upper δ-level curve of a fuzzy a priori density
$\overline{\pi}_\delta(\cdot\|D)$	upper δ-level curve of a fuzzy a-posteriori density
$\underline{\pi}_\delta(\cdot)$	lower δ-level curve of a fuzzy a priori density
$\underline{\pi}_\delta(\cdot\|D)$	lower δ-level curve of a fuzzy a posteriori density
$\pi(\cdot)$	a priori density
$\pi(\cdot\|x_1,\ldots,x_n)$	a posteriori density
$\pi(\cdot\|D)$	a posteriori density
$\pi^*(\cdot)$	fuzzy a priori density
$\pi^*(\cdot\|D^*)$	fuzzy a posteriori density
$\displaystyle\prod_{i=1}^{n}$	product operator
$\propto$	proportional (up to a multiplicative constant)
$\sim$	distributed
$\tau(\Theta)$	set of all transformed parameters
$\tau(\theta)$	transformed parameter
θ_o	true parameter value
θ	parameter
$\tilde{\theta}$	stochastic quantity describing the parameter in Bayesian context
$\times$	symbol for Cartesian product
$\varphi(\cdot)$	membership function of a fuzzy confidence region
$\vartheta(x_1^*,\ldots,x_n^*)$	function of fuzzy observations
$\widehat{\tau(\theta_0)}^*$	fuzzy estimator of $\tau(\theta_0)$
$\xi(.,\ldots,.)$	vector-characterizing function

$\xi_{\pi^*(\theta)}(\cdot)$	characterizing function of $\pi^*(\theta)$
δ	constant in $(0;1]$
#	number of $\cdots$
$\odot$	multiplication operation for fuzzy numbers
$\oplus$	addition operation for fuzzy numbers
$\subseteq$	subset symbol
$(-\infty; x]$	set of all real numbers not greater than x

A2

Solutions to the problems

Chapter 1

(a) Recreation time after an illness, quality of life data, amount of CO_2 released to the environment in 1 year, position of a plane on a radar screen, hardness of a metal, height of a tree.

(b) Stochastic uncertainty is due to variability of quantities, and probability distributions are the mathematical descriptions of this variability. It models the uncertainty of a quantity before it is observed. After the corresponding experiment is performed, the value of the quantity is a number or vector.

   Contrary to this, fuzziness is the imprecision of the outcomes of experiments after the experiment was performed. In the case of one-dimensional stochastic quantities this data uncertainty is described by fuzzy numbers.

(c) The reading of a digital measurement device gives a number with finite many decimals. The resulting number contains no information on the remaining infinite number of decimals which determine a real number. Therefore, precisely speaking, the reported value is not a real number but the set of all real numbers between the real number $\underline{x}$ which is obtained if all remaining decimals are equal to zero, and the real number $\overline{x}$ which is obtained if all remaining decimals are equal to 9. This is an interval $[\underline{x}; \overline{x}]$.

Statistical Methods for Fuzzy Data   Reinhard Viertl
© 2011 John Wiley & Sons, Ltd

(d) All these pictures are color intensity pictures representing valuable information on real phenomena. Since the boundaries of such color intensity pictures are fuzzy these data are examples of fuzzy data.

Chapter 2

(a)

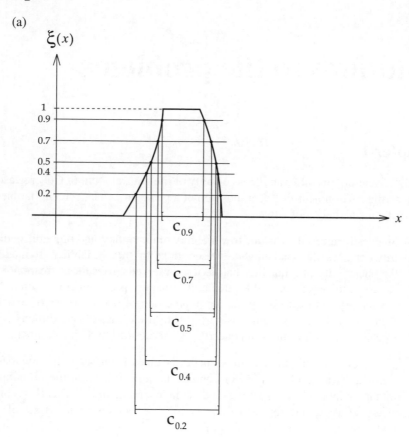

(b)

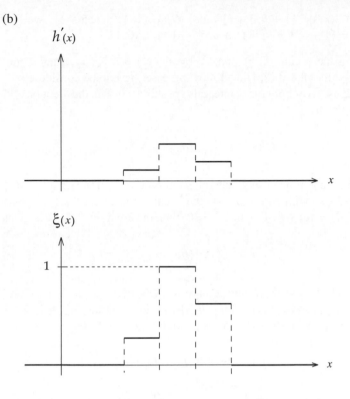

$h'(x)$

$\xi(x)$

1

(c) $C_\delta[\xi(.,.)] = [a_1; b_1] \times [a_2; b_2]$ $\forall \delta \in (0; 1]$.

(d) By Proposition 2.1 the δ-cuts of the combined fuzzy vector are the Cartesian products of the δ-cuts of the fuzzy numbers $\xi_i(\cdot)$, $i = 1(1)n$. Since the δ-cuts of the fuzzy numbers are finite unions of closed intervals, also the Cartesian products are finite unions of so-called rectangles $\overset{n}{\underset{i=1}{\times}} [a_{\delta,j}^{(i)}; b_{\delta,j}^{(i)}]$ which are closed sets.

(e) The δ-cuts of fuzzy intervals are compact intervals of real numbers. Application of the minimum t-norm yields as δ-cuts of the fuzzy combined vector the Cartesian products of the mentioned compact intervals. Therefore the δ-cuts of the combined fuzzy vector are k-dimensional compact intervals.

Chapter 3

(a) The δ-cuts of trapezoidal fuzzy numbers $x^* = t^*(m, s, l, r)$ have the following form: $C_\delta(x^*) = [m - s - l(1 - \delta); m + s + r(1 - \delta)]$ $\forall \delta \in (0; 1]$. Therefore, by applying the minimum t-norm, the δ-cuts of $\underline{x}^*$ are the Cartesian products $C_\delta(x_1^*) \times C_\delta(x_2^*)$.

By $C_\delta(x_1^*) = [1 + \delta; 3 - \delta]$ and $C_\delta(x_2^*) = [1 + \delta; 5 - \delta]$ we obtain $C_\delta(\underline{x}^*) = [1 + \delta; 3 - \delta] \times [1 + \delta; 5 - \delta] \; \forall \delta \in (0; 1].$

(b) Denoting the δ-cuts of x_1^* and x_2^* by $C_\delta(x_1^*) = [a_{\delta,1}; b_{\delta,1}]$ and $C_\delta(x_2^*) = [a_{\delta,2}; b_{\delta,2}]$ for all $\delta \in (0; 1]$ the following cases have to be considered:

(1) If all δ-cuts of both fuzzy intervals are subsets of $\mathbb{R}_+$, the δ-cut of $x_1^* \odot x_2^*$ is given by

$$C_\delta(x_1^* \odot x_2^*) = [a_{\delta,1} \cdot a_{\delta,2}; \; b_{\delta,1} \cdot b_{\delta,2}].$$

(2) If the δ-cut of one fuzzy interval is a subset of $\mathbb{R}_+$, for example $C_\delta(x_1^*) = [a_{\delta,1}; b_{\delta,1}]$ with $a_{\delta,1} \geq 0$, and the δ-cut of x_2^* is a subset of $(-\infty; 0]$, $C_\delta(x_2^*) = [a_{\delta,2}; b_{\delta,2}]$ with $b_{\delta,2} \leq 0$, then the δ-cut of $x_1^* \odot x_2^*$ is given by

$$C_\delta(x_1^* \odot x_2^*) = [b_{\delta,1} \cdot a_{\delta,2}; \; a_{\delta,1} \cdot b_{\delta,2}].$$

(3) If both δ-cuts of x_1^* and x_2^* are subsets of $(-\infty; 0]$ with $C_\delta(x_1^*) = [a_{\delta,1}; b_{\delta,1}]$ and $C_\delta(x_2^*) = [a_{\delta,2}; b_{\delta,2}]$ with $b_{\delta,1} \leq 0$ and $b_{\delta,2} \geq 0$, the δ-cut of the fuzzy product $x_1^* \odot x_2^*$ is given by

$$C_\delta(x_1^* \odot x_2^*) = [b_{\delta,1} \cdot b_{\delta,2}; a_{\delta,1} \cdot a_{\delta,2}].$$

(4) If all δ-cuts of both factors x_1^* and x_2^* contain positive and negative numbers, i.e. $C_\delta(x_1^*) = [a_{\delta,1}; b_{\delta,1}]$ and $C_\delta(x_2^*) = [a_{\delta,2}; b_{\delta,2}]$ with $a_{\delta,1} \leq 0$ and $b_{\delta,1} \geq 0$, $a_{\delta,2} \leq 0$ and $b_{\delta,2} \geq 0$, then we obtain $C_\delta(x_1^* \odot x_2^*) = [\min\{a_{\delta,1} \cdot b_{\delta,2}; \; b_{\delta,1} \cdot a_{\delta,2}\}; \max\{a_{\delta,1} \cdot a_{\delta,2}; \; b_{\delta,1} \cdot b_{\delta,2}\}].$

(5) If the δ-cut of one fuzzy factor, say x_1^*, is a subset of $\mathbb{R}_+$, i.e. $C_\delta(x_1^*) = [a_{\delta,1}; b_{\delta,1}]$ with $a_{\delta,1} \geq 0$, and $C_\delta(x_2^*)$ contains positive and negative numbers, i.e. $C_\delta(x_2^*) = [a_{\delta,2}; b_{\delta,2}]$ with $a_{\delta,2} \leq 0$ and $b_{\delta,2} \geq 0$, then we obtain

$$C_\delta(x_1^* \odot x_2^*) = [b_{\delta,1} \cdot a_{\delta,2}; \; b_{\delta,1} \cdot b_{\delta,2}].$$

(6) If the δ-cut of one factor is a subset of $(-\infty; 0]$, for example $C_\delta(x_1^*) = [a_{\delta,1}; b_{\delta,1}]$ with $b_{\delta,1} \leq 0$, and the δ-cut of x_2^* contains both, negative and positive numbers, i.e. $C_\delta(x_2^*) = [a_{\delta,2}; b_{\delta,2}]$ with $a_{\delta,2} \leq 0$ and $b_{\delta,2} \geq 0$, then we obtain

$$C_\delta(x_1^* \odot x_2^*) = [a_{\delta,1} \cdot b_{\delta,2}; \; a_{\delta,1} \cdot a_{\delta,2}].$$

Summarizing the above six cases we obtain the result given in Remark 3.3.

(c) Obeying that the sum of two trapezoidal fuzzy numbers is a trapezoidal fuzzy number, the characterizing function of $x_1^* \oplus x_2^*$ is the following:

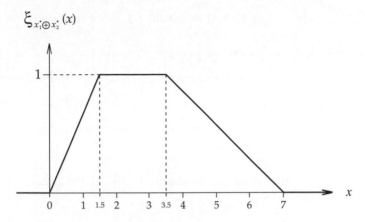

(d) Application of the result of (b) (or Remark 3.3) some δ-cuts are obtained. Drawing the result gives the following characterizing function:

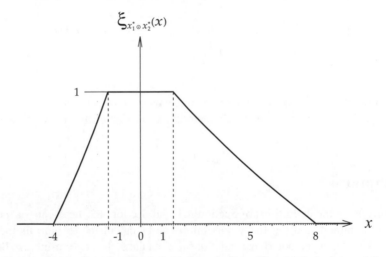

(e) For a classical real valued function $f : \mathbb{R} \to \mathbb{R}$ which is integrable, the characterizing function $\xi_{f(x)}(\cdot)$ of $f(x)$ is the indicator function $I_{\{f(x)\}}(\cdot)$. The δ-cuts of $I_{\{f(x)\}}(\cdot)$ are the one-point set $\{f(x)\} = [f(x); f(x)]$. Therefore all δ-level functions are identical, i.e. $\underline{f}_\delta(x) = \overline{f}_\delta(x) = f(x) \; \forall x \in \mathbb{R}$ and $\forall \delta \in (0; 1]$. Therefore the generalized integral $\mathcal{J}^* = \oint_a^b f(x)\, dx$, obeying

$$\underline{\mathcal{J}}_\delta = \int_a^b \underline{f}_\delta(x)\,dx = \int_a^b \overline{f}_\delta(x)\,dx = \int_a^b f(x)\,dx = \overline{\mathcal{J}}_\delta, \; \mathcal{J}^* \text{ reduces to } \int_a^b f(x)\,dx.$$

(f) Let $f^*(x) \equiv y^*$ with characterizing function $\eta(\cdot)$ for all $x \in [a; b]$. Then the δ-cut $C_\delta[f^*(x)] = [\underline{f}_\delta(x); \overline{f}_\delta(x)] = C_\delta[\eta(\cdot)] \; \forall \delta \in (0; 1]$. Defining $C_\delta[\eta(\cdot)] = [a_\delta; b_\delta] \; \forall \delta \in (0; 1]$ we obtain $\int\limits_a^b \underline{f}_\delta(x)dx = \int\limits_a^b a_\delta \, dx = a_\delta(b-a)$

and $\int\limits_a^b \overline{f}_\delta(x)dx = \int\limits_a^b b_\delta \, dx = b_\delta(b-a) \quad \forall \delta \in (0; 1]$.

Therefore the δ-cuts of the characterizing function $\chi(\cdot)$ of the generalized integral $\oint\limits_a^b f^*(x)dx$ is $[a_\delta(b-a); b_\delta(b-a)] \quad \forall \delta \in (0; 1]$.

The resulting characterizing function $\chi(\cdot)$ is given by its values

$$\chi(x) = \sup\left\{\delta \cdot I_{C_\delta(J^*)}(x) : \delta \in [0; 1]\right\} \; \forall x \in \mathbb{R}.$$

Now $\sup\left\{\delta \cdot I_{C_\delta(J^*)}(x) : \delta \in [0; 1]\right\}$

$$= \sup\left\{\delta \cdot I_{[a_\delta(b-a); b_\delta(b-a)]}(x) : \delta \in [0; 1]\right\}$$
$$= (b-a) \cdot \eta(x) \; \forall x \in \mathbb{R}.$$

By this, the generalized integral is the fuzzy interval whose characterizing function is $(b-a) \cdot \eta(\cdot)$, where $\eta(\cdot)$ is the characterizing function of the fuzzy value $f^*(x)$.

Chapter 4

(a) Using the definitions of Section 4.1 and Section 4.2 the δ-cuts of the fuzzy minimum and the fuzzy maximum can be calculated. From this the corresponding characterizing functions are obtained. The two results are depicted in the following diagram:

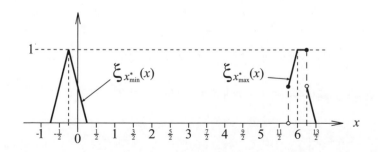

(b)

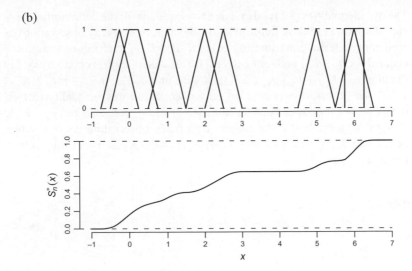

Chapter 5

(a)

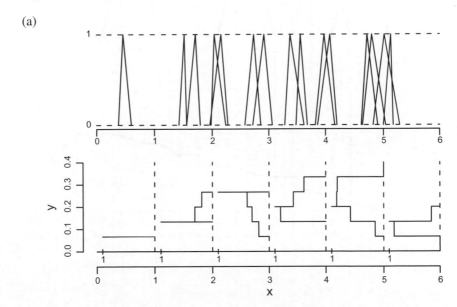

(b) The two inequalities (5.1) and (5.2) are consequences of the definition of the δ-cuts of the fuzzy relative frequencies. If the support of a fuzzy observation has non-empty intersection with two classes K_j and K_{j+1}, then this observation is counted for $\overline{h}_{n,\delta}(K_j)$ and for $\overline{h}_{n,\delta}(K_{j+1})$. This observation is counted twice in the sum $\overline{h}_{n,\delta}(K_j) + \overline{h}_{n,\delta}(K_{j+1})$. If taking the union $K_j \cup K_{j+1}$ for $\overline{h}_{n,\delta}(K_j \cup K_{j+1})$ the observation is counted only once. From this inequality (5.1) is obtained. For the lower boundaries $\underline{h}_{n,\delta}(K_j)$ and $\underline{h}_{n,\delta}(K_{j+1})$, and $\underline{h}_{n,\delta}(K_j \cup K_{j+1})$ it is possible that the support of a fuzzy observation is not a subset of K_j and K_{j+1}, but it can be a subset of $K_j \cup K_{j+1}$. From this follows inequality (5.2).

Chapter 6

(a)

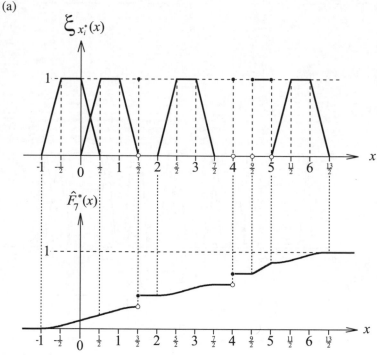

(b)

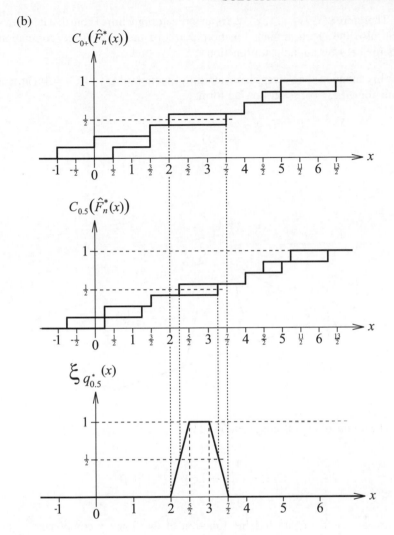

Chapter 7

(a) For data in the form of two-dimensional intervals $(x, y)_i^*$ which has vector-characterizing function $\xi_i(.,.) = I_{[a_{1,i};b_{2,i}]\times[a_{1,i};b_{2,i}]}(.,.)$, $i = 1(1)n$ the vector-characterizing function of the fuzzy combined sample is given by $\xi(x_1, y_1, \ldots, x_n, y_n) = \min\{\xi_i(x_i, y_i) : i = 1(1)n\}$ and the δ-cuts of $\xi(., \ldots, .)$ are the Cartesian products

$$C_\delta[\xi(., \ldots, .)] = \underset{i=1}{\overset{n}{X}} C_\delta[\xi_i(.,.)] = \underset{i=1}{\overset{n}{X}} \left([a_{i,1}; b_{i,1}] \times [a_{i,2}; b_{i,2}]\right)$$

Therefore $\xi(x_1, y_1, \ldots, x_n, y_n)$ is an indicator function. From that it follows that also the characterizing function $\psi_{r^*}(\cdot)$ of the generalized correlation coefficient r^* is an indicator function.

(b) In this situation the vector-characterizing function of $(x, y)_i^*$, $i = 1(1)n$ is a function $\xi_i(x, y)$ of the following form:

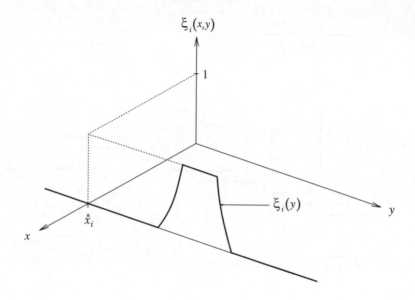

The δ-cuts of $(x, y)_i^*$ are given by

$$C_\delta\left((x, y)_i^*\right) = \{\overset{\circ}{x}_i\} \times C_\delta\left[\xi_i(\cdot)\right]$$

where $\xi_i(\cdot)$ is the characterizing function of the fuzzy y-component y_i^* of $(x, y)_i^*$.

The δ-cuts of the fuzzy combined sample $(x_1, y_2, \ldots, x_n, y_n)^*$ are given by

$$C_\delta\left((x_1, y_1, \ldots, x_n, y_n)^*\right) = \underset{i=1}{\overset{n}{\times}} C_\delta\left((x, y)_i^*\right)$$

$$= \underset{i=1}{\overset{n}{\times}} \left(\{\overset{\circ}{x}_i\} \times C_\delta\left[\xi_i(\cdot)\right]\right).$$

The vector-characterizing function $\xi(x_1, y_1, \ldots, x_n, y_n)$ can be obtained by the construction lemma for membership functions.

Chapter 8

(a) Let the fuzzy values $f^*(x)$ be LR-fuzzy numbers (cf. Section 2.1). Then the δ-cuts are given by

$$C_\delta\left(f^*(x)\right) = \left[m_x - s_x - l_x \cdot L^{-1}(\delta);\ m_x + s_x + r_x \cdot R^{-1}(\delta)\right].$$

Therefore the upper and lower δ-level curves $\overline{f}_\delta(\cdot)$ and $\underline{f}_\delta(\cdot)$ are

$$\left.\begin{aligned}\overline{f}_\delta(x) &= m(x) + s(x) + r(x) \cdot R^{-1}(\delta)\\ \text{and}&\\ \underline{f}_\delta(x) &= m(x) - s(x) - l(x) \cdot L^{-1}(\delta)\end{aligned}\right\} \quad \forall \delta \in (0; 1].$$

The fuzzy integral $\mathcal{J}^* = \int_0^1 f^*(x)dx$ is defined by its δ-cuts $C_\delta(\mathcal{J}^*)$ which

are given by $C_\delta(\mathcal{J}^*) = \left[\int_0^1 \underline{f}_\delta(x)dx;\ \int_0^1 \overline{f}_\delta(x)dx\right].$

For a fuzzy uniform density we can assume $m(x) \equiv 1\ \forall x \in [0; 1]$.
Now we obtain

$$\int_0^1 \overline{f}_\delta(x)dx = \int_0^1 1 + s(x) + r(x) \cdot R^{-1}(\delta)dx$$

and

$$\int_0^1 \underline{f}_\delta(x)dx = \int_0^1 1 - s(x) - l(x) \cdot L^{-1}(\delta)dx.$$

Assuming the functions $s(\cdot)$ and $l(\cdot)$ to be integrable we obtain

$$\int_0^1 \overline{f}_\delta(x)dx = 1 + \int_0^1 s(x)dx + R^{-1}(\delta)\int_0^1 r(x)dx$$

and

$$\int_0^1 \underline{f}_\delta(x)dx = 1 - \int_0^1 s(x)dx - L^{-1}(\delta)\int_0^1 l(x)dx.$$

From this the δ-cuts $C_\delta(\mathcal{J}^*)$ are obtained, and by the representation lemma 2.1 the characterizing function of $\mathcal{J}^*$ is determined.

(b) The fuzzy probability $P^*\left([0;\frac{1}{2}]\right)$ based on the fuzzy density $f^*(\cdot)$ of (a) is obtained via its δ-cuts $[\underline{P}_\delta; \overline{P}_\delta]$.

$$\left.\begin{array}{c}\overline{P}_\delta = \sup\left\{\left[\int_0^{1/2} f(x)dx : f(\cdot) \in S_\delta\right]\right\} \\[2em] \underline{P}_\delta = \inf\left\{\left[\int_0^{1/2} f(x)dx : f(\cdot) \in S_\delta\right]\right\}\end{array}\right\} \quad \forall \delta \in (0; 1]$$

Depending on the functions $\overline{f}_\delta(\cdot)$ and $\underline{f}_\delta(\cdot)$ the values $\overline{P}_\delta$ and $\underline{P}_\delta$ are calculated.

For the special case of constant fuzzy value $f^*(x)$ with trapezoidal characterizing function $t^*(m, s, l, r)$ the calculation of $\underline{P}_\delta$ and $\overline{P}_\delta$ is relatively simple.

Chapter 9

(a) If all δ-cuts of $\psi(\cdot)$ are finite unions of compact intervals then the convex hull of $\psi(\cdot)$ is $\psi(\cdot)$ itself. If some δ-cuts of $\psi(\cdot)$ are not compact or not finite unions of compact intervals then the function $\xi(\cdot)$ is defined by

$$\xi(x) := \sup\{\min\{\psi(x_1), \psi(x_2)\} : \alpha x_1 + (1 - \alpha)x_2 = x\}.$$

A concrete example is the following:

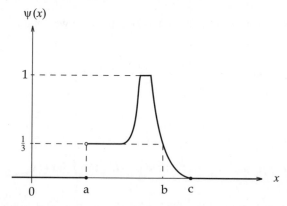

The $\frac{1}{3}$-cut of $\psi(\cdot)$ is $(a; b]$ which is not closed. Therefore the convex hull $\xi(\cdot)$ is the following

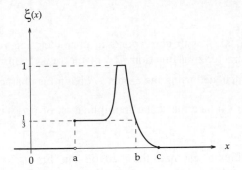

(b) From $\beta < \delta$ it follows $C_\beta(x^*) \supseteq C_\delta(x^*)$ by definition of δ-cuts. Therefore we obtain

$$\bigcap_{\beta<\delta} C_\beta(x^*) \supseteq C_\delta(x^*).$$

In order to prove the equality of both sets assume $x \in C_\beta(x^*)$ for all $\beta > \delta$. Then $\xi_{x^*}(x) \geq \beta$ for all $\beta < \delta$. Assuming $\xi_{x^*}(x) < \delta$ yields a contradiction. Therefore $\bigcap_{\beta<\delta} C_\beta(x^*) \subseteq C_\delta(x^*)$ and the equality follows.

Chapter 10

(a) For sample size one.

(b) For fuzzy intervals x_i^*, $i = 1(1)n$ the δ-cuts are closed finite intervals $[a_{\delta,i}; b_{\delta,i}]$, and the Cartesian products

$$\underset{i=1}{\overset{n}{\times}} C_\delta(x_i^*) = \underset{i=1}{\overset{n}{\times}} [a_{\delta,i}; b_{\delta,i}]$$

are n-dimensional intervals, which are simply connected compact sets. Therefore the combined fuzzy sample is an n-dimensional fuzzy vector.

(c)

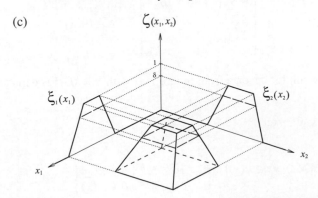

Chapter 11

(a) For n fuzzy observations x_i^* with corresponding characterizing functions $\xi_i(\cdot)$, $i = 1(1)n$ the characterizing function $\mu(\cdot)$ of the fuzzy sample mean $\overline{x}_n^* = \frac{1}{n} \bigoplus_{i=1}^{n} x_i^*$ can be calculated using the δ-cuts to obtain the characterizing function of $\bigoplus_{i=1}^{n} x_i^*$, and then take the scalar multiplication of fuzzy numbers from Section 3.3.

(b) There are two possibilities to estimate the median: The first is to use the smoothed empirical distribution function $\hat{F}_n^{sm}(\cdot)$. If the corresponding 0.5 fractile exists, this can be used as estimator for the median. The second possibility is based on the fuzzy valued empirical distribution function $\hat{F}_n^*(\cdot)$ from Section 6.1. The fuzzy estimate of the median is the fuzzy fractile $q_{0.5}^*$ from Section 6.2.

Chapter 12

(a) The fuzzy confidence region can be constructed in the following way: The membership function $\varphi(\cdot)$ of the fuzzy confidence region (here a fuzzy confidence interval) $\Theta_{1-\alpha}^*$ for θ can be obtained using Proposition 12.1. Taking the confidence function $\kappa(X_1, \ldots, X_n)$ given in Example 12.1, the δ-cuts of $\Theta_{1-\alpha}^*$ are

$$C_\delta(\Theta_{1-\alpha}^*) = \left[\min_{\underline{x} \in C_\delta(\underline{x}^*)} \underline{\kappa}(\underline{x}); \max_{\underline{x} \in C_\delta(\underline{x}^*)} \overline{\kappa}(\underline{x}) \right] \quad \forall \delta \in (0; 1].$$

The minimum is obtained for the left boundaries $a_{\delta,i}$ of the δ-cuts of x_i^*, and the maximum for the right boundaries $b_{\delta,i}$ of the δ-cuts of x_i^*. Therefore we obtain

$$C_\delta(\Theta_{1-\alpha}^*) = \left[\frac{2\sum_{i=1}^{n} a_{\delta,i}}{\chi^2_{2n;1-\frac{\alpha}{2}}}; \frac{2\sum_{i=1}^{n} b_{\delta,i}}{\chi 2n; \frac{\alpha}{2}} \right] \quad \forall \delta \in (0; 1].$$

The numerical result for the data from Figure 12.1 with $\alpha = 0.05$ is depicted in Figure 12.2.

(b) Let $\theta \in \bigcup_{\underline{x}: \xi(\underline{x})=1} \kappa(\underline{x})$, then

$$\varphi(\theta) = \begin{cases} \sup\{\xi(\underline{x}) : \theta \in \kappa(\underline{x})\} & \text{if} \quad \exists \underline{x} \in M^n : \theta \in \kappa(\underline{x}) \\ 0 & \text{if} \quad \nexists \underline{x} \in M^n : \theta \in \kappa(\underline{x}) \end{cases} \quad \forall \theta \in \Theta,$$

and $\exists \underline{x} : \theta \in \kappa(\underline{x})$. Moreover by assumption $\xi(\underline{x}) = 1$, and therefore $\sup\{\xi(\underline{x}) : \theta \in \kappa(\underline{x})\} = 1$ and $\varphi(\theta) = 1$. By $I \bigcup_{\underline{x}:\xi(\underline{x})=1} \kappa(\underline{x})(\theta) = 1$ for $\theta \in$

$\bigcup_{\underline{x}:\xi(\underline{x})=1} \kappa(\underline{x})$ the assertion follows.

Chapter 13

(a) The classical test statistic for the hypothesis $\mathcal{H}_0 : \mu = \mu_0$ is the following:

$$T = \frac{\overline{X}_n - \mu_0}{S_n / \sqrt{n}}.$$

For fuzzy data $x_1^*, \ldots, x_n^*$, using the extension principle the characterizing function $\xi_{t^*}(\cdot)$ of the fuzzy value $t^* = \frac{\overline{x}_n^* - \mu_0}{s_n^* / \sqrt{n}}$ of the test statistic can be obtained. Based on the support of t^* the p-value is determined.

(b) In order to construct the characterizing function of the fuzzy p-value p^* a suitable number of δ-values is chosen and the corresponding δ-cuts of p^* are calculated.

(c) This is done as described before Proposition 13.1.

Chapter 14

(a) Fuzzy discrete probability distributions on finite sets $\{x_1, \ldots, x_n\}$ are given by their fuzzy point probabilities $p^*(x_j)$ which are fuzzy intervals. For a subset $\{x_{j1}, \ldots, x_{jm}\}$ of $\{x_1, \ldots, x_k\}$ the probability $P^*(\{x_{j1}, \ldots, x_{jm}\})$ is given via its δ-cuts $C_\delta \left(P^*(\{x_{j1}, \ldots, x_{jm}\}) \right)$ which are defined by all possible sums of numbers $y_l \in C_\delta(p^*(x_{jl}))$ with

$$\sum_{l=1}^{n} y_l = 1.$$

By this construction condition (1) from Definition 8.3 is fulfilled. Condition (2) follows from the assumption that $\bigoplus_{j=1}^{k} p^*(x_j)$ is a fuzzy interval with

$$1 \in C_1 \left[\bigoplus_{j=1}^{k} p^*(x_j) \right].$$

In order to prove condition (3) consider pair wise disjoint subsets $A_1, \ldots, A_m$ of $\{x_1, \ldots, x_k\}$. Denoting $\bigcup_{i=1}^{m} A_i$ by $\{x_{j1}, \ldots, x_{jq}\}$ the value

$\overline{P}_\delta \left(\bigcup_{i=1}^{m} A_i \right)$ is obtained as the maximum of all sums $\sum_{l=1}^{q} y_l$ with $y_l \in$

$C_\delta \left(p^*(x_{jl}) \right)$ where $\sum_{l=1}^{n} y_l = 1$. On the other hand $\overline{P}_\delta(A_i)$ is obtained as the

maximum of all sums $\sum_{l=1}^{r_i} y_l$ with $y_l \in C_\delta \left(p^*(x_{jl}) \right)$ with $A_i = \{x_{i_1}, \ldots, x_{i_{r_i}}\}$.

But all possible sums of maximums of partial sums are not less than the

first considered maximum of sums. Therefore $\overline{P}_\delta \left(\bigcup_{i=1}^{m} A_i \right) \leq \sum_{i=1}^{m} \overline{P}_\delta(A_i)$. A

similar argument for the minimum of sums $\sum_{l=1}^{q} y_l$ gives the second inequality.

(b) For the upper δ-level curves of the fuzzy a posteriori density conditional on a first sample $\underline{x}_1$ we obtain

$$\overline{\pi}_\delta(\theta|\underline{x}_1) = \frac{\overline{\pi}_\delta(\theta)\overline{l}(\theta;\underline{x}_1)}{\int_\Theta \frac{1}{2}\left[\underline{\pi}_\delta(\theta)l(\theta;\underline{x}_1) + \overline{\pi}_\delta(\theta)\overline{l}(\theta;\underline{x}_1)\right]d\theta}.$$

Using the abbreviation $N(\underline{x}_1)$ from section 14.2 we obtain for a second sample $\underline{x}_2$; taking $\pi^*(\theta|\underline{x}_1)$ as new a priori density:

$$\overline{\pi}_\delta(\theta|\underline{x}_1, \underline{x}_2) = \frac{\overline{\pi}_\delta(\theta|\underline{x}_1) \cdot l(\theta;\underline{x}_2)}{\int_\Theta \frac{\underline{\pi}_\delta(\theta|\underline{x}_1)+\overline{\pi}_\delta(\theta|\underline{x}_1)}{2} l(\theta;\underline{x}_2)d\theta}$$

$$= \frac{N(\underline{x})^{-1} \cdot \overline{\pi}_\delta(\theta) \cdot l(\theta;\underline{x}_1) \cdot l(\theta;\underline{x}_2)}{\int_\Theta N(\underline{x}_1)^{-1} \cdot \frac{\underline{\pi}_\delta(\theta) \cdot l(\theta;\underline{x}_1)+\overline{\pi}_\delta(\theta) \cdot l(\theta;\underline{x}_1)}{2} \cdot l(\theta;\underline{x}_2)d\theta}$$

$$= \frac{\overline{\pi}_\delta(\theta) \cdot l(\theta;\underline{x}_1) \cdot l(\theta;\underline{x}_2)}{\int_\Theta \frac{\underline{\pi}_\delta(\theta)+\overline{\pi}_\delta(\theta)}{2} \cdot l(\theta;\underline{x}_1) \cdot l(\theta;\underline{x}_2)d\theta} = \overline{\pi}_\delta(\theta|\underline{x})$$

where $\underline{x} = (\underline{x}_1, \underline{x}_2)$ is the combined sample.

(c) For classical a priori density $\pi(\cdot)$ the upper and lower δ-level curves coincide, i.e. $\underline{\pi}_\delta(\theta) = \overline{\pi}_\delta(\theta) = \pi(\theta)$ $\forall \theta \in \Theta$. Therefore only one δ-level curve $\pi(\cdot|\underline{x})$ is obtained which is the classical a posteriori density.

Chapter 15

(a)

$$\overline{\pi}_\delta(\theta|\underline{x}_1^*, \underline{x}_2^*) = \frac{\overline{\pi}_\delta(\theta|\underline{x}_1^*) \cdot \overline{l}_\delta(\theta;\underline{x}_2^*)}{\int_\Theta \frac{1}{2}\left[\underline{\pi}_\delta(\theta|\underline{x}_1^*) \cdot \underline{l}_\delta(\theta;\underline{x}_2^*) + \overline{\pi}_\delta(\theta|\underline{x}_1^*) \cdot \overline{l}_\delta(\theta;\underline{x}_2^*)\right]d\theta}$$

$$= \frac{N(\underline{x}_1^*)^{-1} \cdot \overline{\pi}_\delta(\theta) \cdot \overline{l}_\delta(\theta; \underline{x}_1^*) \cdot \overline{l}_\delta(\theta; \underline{x}_2^*)}{\int\limits_\Theta \frac{1}{2} N(\underline{x}_1^*)^{-1} \left[\underline{\pi}_\delta(\theta) \cdot \underline{l}_\delta(\theta; \underline{x}_1^*) \cdot \underline{l}_\delta(\theta; \underline{x}_2^*) + \overline{\pi}_\delta(\theta) \cdot \overline{l}_\delta(\theta; \underline{x}_1^*) \cdot \overline{l}_\delta(\theta; \underline{x}_2^*) \right] d\theta}$$

$$= \frac{\overline{\pi}_\delta(\theta) \cdot \overline{l}_\delta(\theta; \underline{x}_1^*) \cdot \overline{l}_\delta(\theta; \underline{x}_2^*)}{\int\limits_\Theta \frac{1}{2} \left[\underline{\pi}_\delta(\theta) \cdot \underline{l}_\delta(\theta; \underline{x}_1^*) \cdot \underline{l}_\delta(\theta; \underline{x}_2^*) + \overline{\pi}_\delta(\theta) \cdot \overline{l}_\delta(\theta; \underline{x}_1^*) \cdot \overline{l}_\delta(\theta; \underline{x}_2^*) \right] d\theta}$$

$$= \frac{\overline{\pi}_\delta(\theta) \cdot \overline{l}_\delta(\theta; \underline{x}^*)}{\int\limits_\Theta \frac{1}{2} \left[\underline{\pi}_\delta(\theta) \cdot \underline{l}_\delta(\theta; \underline{x}^*) + \overline{\pi}_\delta(\theta) \cdot \overline{l}_\delta(\theta; \underline{x}^*) \right] d\theta} = \overline{\pi}_\delta(\theta | \underline{x}^*)$$

with $\underline{x}^* = (x_1^*, x_2^*)$ combined by the minimum t-norm.

(b) For precise data $\underline{x} = (x_1, \ldots, x_n)$ and standard a priori density $\pi(\cdot)$ of $\tilde{\theta}$ all δ-level functions of $\pi(\cdot)$ are identical with $\pi(\cdot)$. Also the likelihood function is a classical real valued function. Therefore only one δ-level function is obtained by the generalized Bayes' theorem, i.e.

$$\underline{\pi}_\delta(\theta | \underline{x}^*) = \overline{\pi}_\delta(\theta | \underline{x}^*) = \pi(\theta | \underline{x}^*) \quad \forall \theta \in \Theta.$$

Chapter 16

(a) In the case of precise sample $\underline{\mathring{x}} = (\mathring{x}_1, \ldots, \mathring{x}_n)$ the vector-characterizing function of the combined sample is the one-point indicator function $\xi(., \ldots, .) = I_{\{(\mathring{x}_1, \ldots, \mathring{x}_n)\}}(., \ldots, .)$. Therefore the membership function $\varphi(\cdot)$ of the corresponding generalized confidence set $\Theta_{1-\alpha}^*$ has values

$$\varphi(\theta) = \begin{cases} \sup \left\{ I_{\{(\mathring{x}_1, \ldots, \mathring{x}_n)\}}(\underline{x}) : \theta \in \Theta_{\underline{x}, 1-\alpha} \right\} & \text{if} \quad \exists \underline{x} : \theta \in \Theta_{\underline{x}, 1-\alpha} \\ 0 & \text{if} \quad \nexists \underline{x} : \theta \in \Theta_{\underline{x}, 1-\alpha} \end{cases}.$$

Since the indicator function $I_{\{(\mathring{x}_1, \ldots, \mathring{x}_n)\}}(\underline{x})$ equals 1 if and only if $\underline{x} = (\mathring{x}_1, \ldots, \mathring{x}_n)$ we have $\varphi(\theta) = 1$ if and only if $\theta \in \Theta_{\underline{\mathring{x}}, 1-\alpha}$. Therefore $\varphi(\cdot) = I_{\Theta_{\underline{\mathring{x}}, 1-\alpha}}$.

(b) For precise sample $\underline{\mathring{x}} = (\mathring{x}_1, \ldots, \mathring{x}_n)$ the membership function $\psi(\cdot)$ of the generalized HPD-region $\Theta_{H, 1-\alpha}^*$ has values

$$\psi(\theta) = \begin{cases} \sup \left\{ I_{\underline{\mathring{x}}}(\underline{x}) : \int\limits_{B(\theta, \underline{x})} \pi(\theta^1 | \underline{x}) d\theta^1 \leq 1 - \alpha \right\} & \text{if} \quad \exists \underline{x} : \theta \in \Theta_{\underline{x}, 1-\alpha} \\ 0 & \text{otherwise} \end{cases} \quad \forall \theta \in \Theta.$$

Now $\psi(\theta)$ is an indicator function, and $\psi(\theta) = 1$ if and only if $\int\limits_{B(\theta; \underline{\mathring{x}})} \pi(\theta^1 | \underline{\mathring{x}}) d\theta^1 \leq 1 - \alpha$, and $\psi(\theta) = 0$ if and only if $\int\limits_{B(\theta; \underline{\mathring{x}})} \pi(\theta^1 | \underline{\mathring{x}}) d\theta^1 > 1 - \alpha$.

Let $\Theta_{H,1-\alpha}$ be the classical HPD-region based on the precise sample $\overset{\circ}{\underline{x}}$. Then

$$\int_{\Theta_{H,1-\alpha}} \pi(\theta^1 | \overset{\circ}{\underline{x}}) \, d\theta^1 = 1 - \alpha \quad \text{and} \quad \pi(\theta^1 | \overset{\circ}{\underline{x}}) \geq C \;\; \forall \theta^1 \in \Theta_{H,1-\alpha},$$

where C is the maximal constant such that the foregoing equation is fulfilled. For $\theta \in \Theta_{H,1-\alpha}$ it follows $\int_{B(\theta;\overset{\circ}{\underline{x}})} \pi(\theta^1 | \overset{\circ}{\underline{x}}) d\theta^1 \leq 1 - \alpha$ and therefore $\psi(\theta) = 1$.

For $\theta \notin \Theta_{H,1-\alpha}$ we have $\int_{B(\theta;\overset{\circ}{\underline{x}})} \pi(\theta^1 | \overset{\circ}{\underline{x}}) d\theta^1 > 1 - \alpha$ and therefore $\psi(\theta) = 0$.

Chapter 17

(a) For discrete quantity $X \sim p(x|\theta)$, $\theta \in \Theta$ with finite observation space $M = \{x_1, \ldots, x_m\}$ and continuous parameter space Θ, for fuzzy a posteriori density $\pi^*(\cdot | D^*)$ the fuzzy numbers $p^*(x_j | D^*) = \displaystyle\oint_{\Theta} p(x_j | \theta) \cdot \pi^*(\theta | D^*) d\theta$ are fulfilling the following. Since $p(\cdot | \theta)$ is a classical discrete probability distribution and $\pi^*(\cdot | D^*)$ a fuzzy probability density, the fuzzy values $p^*(x_j | D^*) = \displaystyle\oint_{\Theta} p(x_j | \theta) \cdot \pi^*(\theta | D^*) d\theta$ have δ-cuts $C_\delta[p^*(x_j | D^*)] = [\underline{p}_\delta(x_j); \overline{p}_\delta(x_j)]$ with

$$\left.\begin{aligned}
\overline{p}_\delta(x_j) &= \int_\Theta p(x_j | \theta) \cdot \overline{\pi}_\delta(\theta | D^*) d\theta \\
\text{and} \quad & \\
\underline{p}_\delta(x_j) &= \int_\Theta p(x_j | \theta) \cdot \underline{\pi}_\delta(\theta | D^*) d\theta
\end{aligned}\right\} \quad \forall \delta \in (0; 1].$$

The fuzzy sum $\displaystyle\bigoplus_{j=1}^{m} p^*(x_j | D^*)$ is a fuzzy interval whose characterizing function has δ-cuts $\left[\displaystyle\sum_{j=1}^{m} \underline{p}_\delta(x_j); \sum_{j=1}^{m} \overline{p}_\delta(x_j)\right]$, and

$$C_1\left(\bigoplus_{j=1}^{m} p^*(x_j | D^*)\right) = \left[\sum_{j=1}^{m} \int_\Theta p(x_j | \theta) \cdot \underline{\pi}_1(\theta | D^*) d\theta; \sum_{j=1}^{m} \int_\Theta p(x_j | \theta) \cdot \overline{\pi}_1(\theta | D^*) d\theta\right].$$

By $\displaystyle\sum_{j=1}^{m} p(x_j | \theta) = 1 \;\; \forall \theta \in \Theta$ and $\pi^*(\cdot | D^*)$ is a fuzzy probability density on Θ, $1 \in C_1\left(\displaystyle\bigoplus_{j=1}^{m} p^*(x_j | D^*)\right)$ and by Section 14.1 $p^*(\cdot | D^*)$ generates a fuzzy discrete distribution on M.

(b) In order to prove that $f^*(\cdot|D^*)$ is a fuzzy probability density we have to prove the existence of a classical probability density $g(\cdot)$ on M_X which fulfills

$$\underline{f}_1(x|D^*) \leq g(x) \leq \overline{f}_1(x|D^*) \quad \forall\, x \in M_X.$$

$\int_\Theta f(x|\theta) \odot \pi^*(\theta|D^*)d\theta$ is calculated using δ-level functions $f(x|\theta) \cdot \overline{\pi}_\delta(\theta|D^*)$ and $f(x|\theta) \cdot \underline{\pi}_\delta(\theta|D^*)$, i.e.

$$\overline{f}_\delta(x|D^*) = \int_\Theta f(x|\theta)\overline{\pi}_\delta(\theta|D^*)\,d\theta$$

and

$$\underline{f}_\delta(x|D^*) = \int_\Theta f(x|\theta)\underline{\pi}_\delta(\theta|D^*)\,d\theta.$$

For $\delta = 1$ we obtain

$$\overline{f}_1(x|D^*) = \int_\Theta f(x|\theta)\overline{\pi}_1(\theta|D^*)\,d\theta$$

and

$$\underline{f}_1(x|D^*) = \int_\Theta f(x|\theta)\underline{\pi}_1(\theta|D^*)\,d\theta.$$

Since $f(\cdot|\theta)$ is a classical probability density and $\pi^*(\cdot|D^*)$ is a fuzzy probability density, there exists a probability density $h(\cdot)$ on Θ with

$$\underline{\pi}_1(\theta|D^*) \leq h(\theta) \leq \overline{\pi}_1(\theta|D^*) \quad \forall\, \theta \in \Theta.$$

Therefore $f(x|\theta) \cdot h(\theta)$ is a two-dimensional density on $M_X \times \Theta$, and $\int_\Theta f(x|\theta)h(\theta)\,d\theta$ as marginal density is a probability density on M_X for which

$$\underline{f}_1(x|D^*) \leq \int_\Theta f(x|\theta)h(\theta)\,d\theta \leq \overline{f}_1(x|D^*) \quad \forall\, x \in M_X.$$

This proves the assertion.

Chapter 18

(a) The classical situation is characterized by the indicator function $I_{U(\theta,d)}(\cdot)$ of $U(\theta, d)$ and a classical probability distribution P on Θ. Therefore the δ-level functions $\underline{U}_\delta(\cdot, d)$ and $\overline{U}_\delta(\cdot, d)$ are all identical and equal to $U(\cdot, d)$. The characterizing function of the generalized expected utility is the indicator function $I_{\mathbb{E}_P U(\theta,d)}(\cdot)$.

(b) If there exists a decision d_B such that the support of the characterizing function of $\mathbb{E}_{P^*} U^*(\tilde{\theta}, d_B)$ has empty intersection with the supports of the characterizing functions $\chi_d(\cdot)$ of the fuzzy expected utilities of all other decisions d, and the left end of $\mathrm{supp}[\chi_{d_B}(\cdot)]$ is greater than all values in all supports of all other decisions, then d_B is the uniquely determined Bayesian decision.

Chapter 19

(a) Polynomial regression functions $\psi(x, \theta_0, \ldots, \theta_k) = \sum_{j=0}^{k} \theta_j x^j$ are linear with respect to the parameters $\theta_0, \ldots, \theta_k$.

Taking $x_j = x^j$ this kind of regression models can be written in the following form:

$$\overset{\circ}{x} = \left(1, x, x^2, \ldots, x^k\right), \quad \boldsymbol{\theta} = (\theta_0, \theta_1, \ldots, \theta_k)$$
$$\psi(x, \theta_0, \ldots, \theta_k) = \boldsymbol{\theta} \overset{\circ}{x}^{\mathrm{T}}$$

This is a special form of the linear regression model (19.4).

(b) The classical linear regression function can be written as general linear model based on the following functions $f_j(\cdot)$:

$$f(\cdot) = \begin{pmatrix} x_1 \\ \vdots \\ x_k \end{pmatrix}, \quad \overset{\circ}{f}(\cdot) = \begin{pmatrix} 1 \\ x_1 \\ \vdots \\ x_k \end{pmatrix},$$
$$\boldsymbol{\theta} = (\theta_0, \theta_1, \ldots, \theta_k)$$

(c) Assumption (3) in Theorem 19.1 implies that $Y_{x_i}, i = 1\,(1)\,n$ are independent.

Chapter 20

(a) First the fuzzy quantities x_1^*, x_2^*, y_1^*, and y_2^* have to be combined into a fuzzy vector with vector-characterizing function

$$\tau(x_1, x_2, y_1, y_2) = \min \{\xi_1(x_1), \xi_2(x_2), \eta_1(y_1), \eta_2(y_2)\}$$
$$\forall (x_1, x_2, y_1, y_2) \in \mathbb{R}^4.$$

Based on this fuzzy combined vector the estimatiors

$$\hat{\theta}_0 = \frac{\left(x_1^2 + x_2^2\right)(y_1 + y_2) - (x_1 + x_2)(x_1 y_1 + x_2 y_2)}{2\left(x_1^2 + x_2^2\right) - (x_1 + x_2)^2}$$

$$\hat{\theta}_1 = \frac{2(x_1 y_1 + x_2 y_2) - (x_1 + x_2)(y_1 + y_2)}{2\left(x_1^2 + x_2^2\right) - (x_1 + x_2)^2}$$

are generalized to the fuzzy estimatiors $\hat{\theta}_0^*$ and $\hat{\theta}_1^*$ by applying the extension principle. For details compare the characterizing functions $\phi_j(\cdot)$ in Remark 20.3 ad (d).

(b) Here the fuzzy elements $\hat{\theta}_0^*$, $\hat{\theta}_1^*$, and x^* with corresponding membership functions have to be combined into a fuzzy 3-dimensional vector z^*. Then the extension principle has to be applied to the function $g(x) = \theta_0 + \theta_1 \cdot x$.

Chapter 21

(a) If the density is given by $g(.| \psi(\theta, \mathbf{x}_i))$ the likelihood function for observed data $(\mathbf{x}_1, y_1), \ldots, (\mathbf{x}_n, y_n)$ is given by

$$l(\theta_0, \ldots, \theta_k; (\mathbf{x}_1, y_1), \ldots, (\mathbf{x}_n, y_n)) = \prod_{i=1}^{n} f_{x_i}(y_i \mid \theta) = \prod_{i=1}^{n} g(y_i \mid \psi(\theta, \mathbf{x}_i)).$$

Bayes' theorem obtains the following form:

$$\pi(\theta_0, \ldots, \theta_k \mid (\mathbf{x}_1, y_i), \ldots, (\mathbf{x}_n, y_n)) \propto$$
$$\pi(\theta_0, \ldots, \theta_k) \prod_{i=1}^{n} g(y_i \mid \psi(\theta_0, \ldots, \theta_k, x_i)).$$

(b)
$$l(\theta_0, \theta_1, (x_1, y_1), \ldots, (x_n, y_n)) = \prod_{i=1}^{n} f_{x_i}(y_i \mid \tau = \theta_0 + \theta_1 x)$$

$$= \prod_{i=1}^{n} \frac{1}{\theta_0 + \theta_1 x_i} \exp\left\{-\frac{y_i}{\theta_0 + \theta_1 x_i}\right\} I_{(0,\infty)}(y_i)$$

$$= \exp\left\{-\sum_{i=1}^{n} \frac{y_i}{\theta_0 + \theta_1 x_i}\right\} \prod_{i=1}^{n} \frac{1}{\theta_0 + \theta_1 x_i} I_{(0,\infty)}(y_i).$$

Bayes' theorem in this case reads

$$\pi\left(\theta_0, \theta_1 \mid (x_1, y_1), \ldots, (x_n, y_n)\right) \propto \quad \pi\left(\theta_0, \theta_1\right) \cdot \exp\left\{-\sum_{i=1}^{n} \frac{y_i}{\theta_0 + \theta_1 x_i}\right\}$$

$$\times \prod_{i=1}^{n} \frac{1}{\theta_0 + \theta_1 x_i} I_{(0,\infty)}(y_i)$$

Chapter 22

(a) The fuzzy valued function

$$\underset{\mathbb{R}^{r-1}}{\int} \pi^*(x_1, \ldots, x_r) \, dx_1 \ldots dx_{s-1} dx_{s+1} \ldots dx_r \text{ has } \delta\text{-level functions}$$

$$\bar{f}_\delta(x_s) = \int_{\mathbb{R}^{r-1}} \bar{\pi}_\delta(x_1, \ldots, x_r) \, dx_1 \ldots dx_{s-1} dx_{s+1} \ldots dx_r \geq 0.$$

By assumption the integral $\int_{\mathbb{R}} \bar{f}_\delta(x_s) \, dx_s$ exists and is finite for all $\delta \in (0; 1]$. Therefore also

$$0 \leq \int_{\mathbb{R}} \underline{f}_\delta(x_s) \, dx_s < \infty \quad \forall \delta \in (0; 1].$$

Now the fuzzy marginal density $f^*(\cdot)$ is defined by its δ-level functions $\underline{f}_\delta(\cdot)$ and $\bar{f}_\delta(\cdot)$ respectively. Furthermore by $\int_{\mathbb{R}} f^*(x) \, dx = \int_{\mathbb{R}^r} \pi^*(x_1, \ldots, x_r) \, dx_1, \ldots dx_r$ it follows $1 \in \text{supp}\left(\int_{\mathbb{R}} f^*(x) \, dx\right)$. Therefore $f^*(\cdot)$ is a fuzzy density.

(b) Let $f(\cdot)$ be a classical probability density with existing expectation, i.e. $\mu = \int_{\mathbb{R}} x f(x) \, dx < \infty$. The δ-level functions of this density are all equal to $f(\cdot)$. Using the definition of Section 22.1 for the expectation we obtain

$$\mu^* = \int_{\mathbb{R}} x \, f(x) \, dx = \mu.$$

The characterizing function $\psi(\cdot)$ of μ^* is given by the family of single point sets $A_\delta = \left\{\int_{\mathbb{R}} x \, f(x) \, dx\right\} = \{\mu\} \quad \forall \, \delta \in (0; 1]$. Therefore we obtain $\psi(\cdot) = I_{\{\mu\}}(\cdot)$.

A3

Glossary

Bayes' formula:	Rule for the calculation of probabilities of hypotheses conditional on observed events.
Bayesian inference:	Statistical inference method where all unknown quantities are modeled by stochastic quantities.
Bayes' theorem:	Rule for the calculation of probability densities of continuous parameters in stochastic models conditional on observed data of continuous variables.
Characterizing function:	Mathematical description of a fuzzy number.
Complete sample:	If n copies of a stochastic quantity are observed and all n observations are obtained then the sample is called complete.
Fuzzy Bayesian inference:	The generalization of Bayesian inference to the situation of fuzzy a priori information and fuzzy data.
Fuzzy combined sample:	Combination of a sample of fuzzy numbers to form a fuzzy element of the sample space.
Fuzzy confidence region:	Generalization of the concept of confidence region for the situation of fuzzy data.
Fuzzy data:	Data which are not precise numbers or classical vectors.
Fuzzy estimate:	Generalized statistical estimation technique for the situation of fuzzy data.
Fuzzy histogram:	A generalized histogram based on fuzzy data.
Fuzzy HPD-regions:	Generalization of highest probability density regions for parameters in Bayesian inference.

Fuzzy information:	Information which is not given in the form of precise numbers, precise vectors, precise functions or precisely defined models.
Fuzzy number:	Mathematical description of a fuzzy one-dimensional quantity.
Fuzzy probability density:	Generalization of a probability density whose values are fuzzy intervals.
Fuzzy probability distribution:	Generalized probability distributions assigning fuzzy intervals to events.
Fuzzy sample:	Data in the form of fuzzy numbers or fuzzy vectors.
Fuzzy statistics:	Statistical analysis methods for the situation of fuzzy information.
Fuzzy utility:	Generalized utility functions assuming fuzzy intervals as values.
Fuzzy valued function:	Generalized real function whose values are fuzzy numbers.
Fuzzy vector:	Mathematical description of a nonprecise (fuzzy) vector quantity.
Gamma function:	Extension of the factorial function to real arguments $x > 0$.
Hybrid data analysis methods:	Data analysis methods using methods from statistics and fuzzy models.
Nonprecise data:	Synonym for fuzzy data.
Triangular norm:	Mathematical operation to combine fuzzy numbers into a fuzzy vector.
Vector-characterizing function:	Mathematical description of a fuzzy vector.

A4

Related literature

There are also other approaches to the analysis of fuzzy data, especially to statistical inference based on fuzzy data. Some are more theoretical oriented, others are centered on special statistical methods like regression analysis, time series analysis, quality control, reliability, and decision analysis. Related books and research papers are given in this Appendix.

The statistical analysis of fuzzy data is an active field of research. Therefore the literature given here cannot be complete. However, based on the references given in the text and the following list of related literature it should be possible for the reader to become acquainted with mathematical methods for the description of fuzzy data and statistical methods for the analysis of fuzzy data.

Arefi, M., Viertl, R. and Taheri, S. (to appear). Fuzzy density estimation, *Metrika*.

Arnold, B. and Gerke, O. (2003) Testing fuzzy linear hypotheses in linear regression, *Metrika*, **57**, 81–95.

Bandemer, H. (ed.) (1993) *Modelling Uncertain Data*, Akademie Verlag, Berlin.

Bandemer, H. (1996) Specifying fuzzy data from grey-tone pictures for pattern recognition, *Pattern Recogn. Lett.*, **17**, 585–592.

Bandemer, H. (2006) *Mathematics of Uncertainty – Ideas, Methods, Application Problems*, Springer, Berlin.

Bertoluzza, C., Gil, M. and Ralescu, D. (eds) (2002) *Statistical Modeling, Analysis and Management of Fuzzy Data*, Physica, Heidelberg.

Bodjanova, S. (2000) A generalized histogram, *Fuzzy Sets Syst.*, **116**, 155–166.

Bodjanova, S. and Viertl, R. (1998) Calculation of Integrals of Fuzzy Functions, Research Report RIS-1998-2, Institut für Statistik und Wahrscheinlichkeitstheorie, Vienna University of Technology, Vienna.

Buckley, J. (2004) *Fuzzy Statistics*, Springer, Berlin.

Cai, K.-Y. (1996) *Introduction to Fuzzy Reliability*, Kluwer, Boston.

Statistical Methods for Fuzzy Data   Reinhard Viertl
© 2011 John Wiley & Sons, Ltd

Colubi, A., García, C. F. and Gil, M. A. (2002) Simulation of random fuzzy variables, *IEEE Trans. Fuzzy Syst.*, **10**, 384–390.

Diamond, P. (1988) Fuzzy least squares, *Inform. Sci.*, **46**, 141–157.

Dubois, D. and Prade, H. (1986) Fuzzy sets and statistical data, *Eur. J. Oper. Res.*, **25**, 345–356.

Dubois, D., Lubiano,M., Prade, H., Gil, M., Grzegorzewski, P. and Hryniewicz, O. (eds) (2008) *Soft Methods for Handling Variability and Imprecision*, Springer, Berlin.

Feng, Y. (1999) Convergence theorems for fuzzy random variables and fuzzy martingales, *Fuzzy Sets Syst.*, **103**, 435–441.

Feng, Y. (2000) Gaussian fuzzy random variables, *Fuzzy Sets Syst.*, **111**, 325–330.

Frühwirth-Schnatter, S. (1993) On fuzzy Bayesian inference, *Fuzzy Sets Syst.*, **60**, 41–58.

Gil, M., Corral, N. and Gil, P. (1988) The minimum inaccuracy estimates in χ^2-tests for goodness of fit with fuzzy observations, *J. Statist. Plann. Inference*, **19**, 95–115.

Gonzales-Rodriguez, G., Blanco, A., Corral, N. and Colubi, A. (2007) Least squares estimation of linear regression models for convex compact random sets, *Adv. Data Anal. Classif.* **1**, 67–81.

Gonzales-Rodriguez, G. Colubi, A. and Trutschnig, W. (2008) Simulating Random Upper Semicontinuous Functions: The Case of Fuzzy Data, Research Report SM-2008-2, Institute of Statistics and Probability Theory, Vienna University of Technology, Vienna.

Grzegorzewski, P. (1998) Statistical inference about the median from vague data, *Control Cybernet.*, **27**, 447–464.

Grzegorzewski, P., Hryniewicz, O. and Gil, M. (eds) (2002) *Soft Methods in Probability, Statistics and Data Analysis*, Physica, Heidelberg.

Höppner, J. (1994) Statistische Prozesskontrolle mit Fuzzy Daten, Doctoral dissertation, University of Ulm, Ulm.

Kacprzyk, J. and Fedrizzi, M. (eds) (1988) *Combining Fuzzy Imprecision with Probabilistic Uncertainty in Decision Making*, Springer Lecture Notes in Economics and Mathematical Systems, Vol. 310, Springer, Berlin.

Klir, G. and Yuan, B. (1995) *Fuzzy Sets and Fuzzy Logic – Theory and Applications*, Prentice Hall, Upper Saddle River.

Kruse, R. and Meyer, K. (1987) *Statistics with Vague Data*, Reidel, Dordrecht.

Kwakernaak, H. (1978) Fuzzy random variables – I. Definition and theorems, *Inform. Sci.*, **15**, 1–29.

Kwakernaak, H. (1979) Fuzzy random variables – II. Algorithms and examples, *Inform. Sci.*, **17**, 253–278.

Manton, K., Woodbury, M. and Tolley, H. (1994) *Statistical Applications Using Fuzzy Sets*, John Wiley & Sons, Ltd, New York.

Ming, M. (1993) On embedding problems of fuzzy number space: Part 5, *Fuzzy Sets Syst.*, **55**, 313–318.

Ming, M. and Cong-Xin, W. (1991) Embedding problem of fuzzy number space: Part 1, *Fuzzy Sets Syst.*, **44**, 33–38.

Miyamoto, S. and Augusta, Y. (1997) L_1 based fuzzy c-means for data with uncertainties, *Comput. Sci. Stat.*, **29**, 408–414.

Möller, B. and Beer, M. (2004) *Fuzzy Randomness – Uncertainty in Civil Engineering and Computational Mechanics*, Springer, Berlin.

Möller, B. and Reuter, U. (2007) *Uncertainty Forecasting in Engineering*, Springer, Berlin, 2007.

Nguyen, H. and Wu, B. (2006) *Fundamentals of Statistics with Fuzzy Data*, Springer, Berlin.

Niculescu, S. and Viertl, R. (1992) A comparison between two fuzzy estimators for the mean, *Fuzzy Sets Syst.*, **48**, 341–350.

Onisawa, T. and Kacprzyk, J. (eds) (1995) *Reliability and Safety Analyses under Fuzziness*, Physica, Heidelberg.

Puri, M. L. and Ralescu, D. A. (1991) Convergence theorem for fuzzy martingales, *J. Math. Anal. Appl.*, **160**, 107–122.

Radström, H. (1952) An embedding theorem for spaces of convex sets, *Proc. Am. Math. Soc.*, **3**, 165–169.

Ross, T., Booker, J. and Parkinson, W. (eds) (2002) *Fuzzy Logic and Probability Applications – Bridging the Gap*, American Statististics Association and SIAM, Philadelphia.

Saade, J. (1994) Extension of fuzzy hypothesis testing with hybrid data, *Fuzzy Sets Syst.*, **63**, 57–71.

Sakawa, M. and Yano, H. (1992) Fuzzy linear regression analysis for fuzzy input-output data, *Inform. Sci.*, **63**, 191–206.

Salicone, S. (2007) *Measurement Uncertainty*, Springer, New York.

Schnatter, S. (1992) Integration-based Kalman-filtering for a dynamic generalized linear trend model, *Comput. Stat. Data Anal.*, **13**, 447–459.

Seising, R. (ed.) (1999) *Fuzzy Theorie und Stochastik – Modelle und Anwendungen in der Diskussion*, Vieweg, Braunschweig.

Slowinski, R. (ed.) (1998) *Fuzzy Sets in Decision Analysis, Operations Research, and Statistics*, Kluwer, Boston.

Song, Q. and Chissom, B. S. (1993) Fuzzy time series and its models, *Fuzzy Sets Syst.*, **54**, 269–277.

Talasová, J. (2003) *Fuzzy Metody*, Univerzita Palackáho v Olomouci, Olomouc.

Trutschnig, W. (2008) A strong consistency result for fuzzy relative frequencies interpreted as estimator for the fuzzy-valued probability, *Fuzzy Sets Syst.*, **159**, 259–269.

Tseng, F.-M. and Tzeng, G.-H. (2002) A fuzzy sesonal ARIMA model for forecasting, *Fuzzy Sets Syst.*, **126**, 367–376.

Viertl, R. (1999) Statistics and integration of fuzzy functions, *Environmetrics* **10**, 487–491.

Viertl, R. (2004) Accelerated life testing, fuzzy information and generalized probability, in M. Nikouline *et al.* (eds) *Parametric and Semiparametric Models with Applications to Reliability, Survival Analysis, and Quality of Life*, Birkhäuser, Boston, pp. 99–105.

Viertl, R. (2006) Univariate statistical analysis with fuzzy data, *Comput. Stat. Data Anal.*, **51**, 133–147.

Viertl, R. (2008a) Foundation of fuzzy Bayesian inference, *J. Uncertain Syst.*, **2**, 187–191.

Viertl, R. (2008b) Fuzzy models for precision measurements, *Mathemat. Comput. Simulat*, **79**, 874–878.

Viertl, R. (2009) On reliability estimation based on fuzzy lifetime data, *J. Statist. Plann. Inference*, **139**, 1750–1755.

Viertl, R. (2010) Degradation and fuzzy information, in M. Nikulin *et al.* (eds) *Advances in Degradation Modeling*, Birkhäuser, Boston, pp. 235–240.

Viertl, R. (to appear) Statistical methods for non-precise data, in M. Lovric (ed.) *International Encyclopedia of Statistical Science*, Springer, Berlin.

Viertl, R. and Hareter, D. (2004a) Fuzzy information and imprecise probability, *Zeitschr. Angew. Math. Mech.*, **84**, 731–739.

Viertl, R. and Hareter, D. (2004b) Generalized Bayes' theorem for non-precise a-priori distribution, *Metrika*, **59**, 263–273.

Wu, H.-C. (2005) Statistical hypotheses testing for fuzzy data, *Inform. Sci.*, **175**, 30–56.

References

Arnold, B. F. (1996) An approach to fuzzy hypothesis testing, *Metrika*, **44**, 119–126.

Catlin, D. E. (1989) *Estimation, Control and the Discrete Kalman Filter*, Springer, New York.

Diamond, P. and Kloeden, P. (1990) Metric spaces of fuzzy sets, *Fuzzy Sets Syst.*, **35**, 241–249.

Diamond, P. and Kloeden, P. (1992) Metric spaces of fuzzy sets: corrigendum, *Fuzzy Sets Syst.*, **45**, 123.

Diamond, P. and Kloeden, P. (1994) *Metric Spaces of Fuzzy Sets*, World Scientific, Singapore.

Dutter, R. and Viertl, R. (1998) Computation of confidence regions based on non-precise data, in *COMPSTAT-Proceedings in Computational Statistics, Short Communications*, IACR, Rothamsted, pp. 27–28.

Feng, Y. (2001) Sums of independent fuzzy random variables, *Fuzzy Sets Syst.*, **123**, 11–18.

Feng, Y., Hu, L. and Shu, H. (2001) The variance and covariance of fuzzy random variables and their application, *Fuzzy Sets Syst.*, **120**, 487–497.

Filzmoser, P. and Viertl, R. (2004) Testing hypotheses with fuzzy data, *Metrika*, **59**, 21–29.

Fréchet, M. (1948) Les éléments aléatoires de natures quelconque dans un éspace distancié, *Ann. Inst. H. Poincaré*, **10**, 215–310.

Gebhardt, J., Gil, M. and Kruse, R. (1997) Fuzzy set-theoretic methods in statistics, in R. Slowinski (ed.) *Handbook of Fuzzy Sets*, Kluwer, Dordrecht, pp. 311–347.

Grewal, M. S. and Andrews, A. P. (2001) *Kalman Filtering*, John Wiley & Sons, Ltd, New York.

Grzegorzewski, P. (2001) Fuzzy tests – defuzzification and randomization, *Fuzzy Sets Syst.*, **180**, 437–446.

Hareter, D. (2003) Zeitreihenanalyse mit unscharfen Daten, Doctoral dissertation, Vienna University of Technology, Vienna.

Janacek, G. (2001) *Practical Time Series*, Oxford University Press, New York.

Klement, E., Mesiar, R. and Pap, E. (2000) *Triangular Norms*, Kluwer, Dordrecht.

Körner, R. (1997a) Linear Models with Random Fuzzy Variables, Dissertation, TU Bergakademie Freiburg.

Körner, R. (1997b) On the variance of random fuzzy variables, *Fuzzy Sets Syst.*, **92**, 83–93.

Körner, R. (2000) An asymptotic α-test for the expectation of random fuzzy variables. *J. Statist. Plann. Inference*, **83**, 331–346.

Körner, R. and Näther, W. (1998) Linear regression with random fuzzy variables: extended classical estimates, best linear estimates, least squares estimates, *Inform. Sci.*, **109**, 95–118.

Körner, R. and Näther, W. (2002) On the variance of random fuzzy variables, *Stud. Fuzziness Soft Comput.*, **87**, 25–42.

Krätschmer, V. (2001) Some complete metrics on spaces of fuzzy subsets, *Fuzzy Sets Syst.*, **130**, 357–365.

Luptáčik, M. (1981) *Nichtlineare Programmierung mit ökonomischen Anwendungen*, Athenäum Verlag, Königstein.

Meinhold, R. J. and Singpurwalla, N. D. (1983) Understanding the Kalman Filter, *Am. Stat.*, **37**, 123–127.

Menger, K. (1951) Ensembles flous et fonctions aleatoires, *C. R. Acad. Sci.*, **232**, 2001–2003.

Munk, K. (1998) Regressionsrechnung mit unscharfen Daten, Diploma thesis, Institute of Statistics and Probability Theory, Vienna University of Technology, Vienna.

Näther, W. (1997) Linear statistical inference for random fuzzy data, *Statistics*, **29**, 221–240.

Näther, W. and Albrecht, M. (1990) Linear regression with random fuzzy observations, *Statistics*, **21**, 521–531.

Parchami, A., Taheri, S. and Mashinchi, M. (2010) Fuzzy *p*-value in testing fuzzy hypotheses with crisp data, *Stat. Papers*, **51**, 209–226.

Puri, M. L. and Ralescu, D. A. (1986) Fuzzy random variables, *J. Math. Anal. Appl.*, **114**, 409–422.

Römer, C. and Kandel, A. Statistical tests for fuzzy data, *Fuzzy Sets Syst.*, **72**, 1–26.

Schlittgen, R. and Streitberg, H. J. (1984) *Zeitreihenanalyse*, 8th edition, Oldenbourg Verlag, Munich.

Schönfeld, P. (1969) *Methoden der Ökonometrie. Band I – Lineare Regressionsmodelle*, Vahlen, Berlin.

Schnatter, S. (1990) Linear dynamic systems and fuzzy data, in R. Trappl (ed.) *Cybernetics and Systems '90*, World Scientific, Singapore.

Taheri, S. (2003) Trends in fuzzy statistics, *Aust. J. Stat.*, **32**, 239–257.

Trutschnig, W. (2006) Fuzzy Probability Distributions, Doctoral dissertation, Vienna University of Technology, Vienna.

Viertl, R. (1987) Is it necessary to develop a fuzzy Bayesian inference, in R. Viertl (ed.) *Probability and Bayesian Statistics*, Plenum Press, New York, pp. 471–475.

Viertl, R. (2002) On the description and analysis of measurement results of continuous quantities, *Kybernetika*, **38**, 353–362.

Viertl, R. and Hareter, D. (2006) *Beschreibung und Analyse unscharfer Information – Statistische Methoden für unscharfe Daten*, Springer, Vienna.

Zadeh, L. A. (1965) Fuzzy sets, *Inform. Control*, **8**, 338–353.

Index